PRAISE FOR *THE MOON*

"The Moon is such a fascinating object in human history. From its dominance in our neighboring sphere, to its romantic impetus, to its symbol of national stature, to its position as a mere stepping stone to greater exploits, the Moon has been omnipresent in the human sky. And no book about the Moon that I've ever read captures the multifaceted nature of the Moon as does Oliver Morton's *The Moon*. As impressive as the subject matter he deals with is the quality of Oliver Morton's writing. Whether describing the clockwork of the lunar cycle or the history of the Moon's influence in human affairs, Morton's writing is clear, incredibly informative, and flows like poetry. A truly delightful and informative read."

—RUSTY SCHWEICKART, lunar module pilot, *Apollo IX*

"A multidisciplinary triumph, combining a deep understanding of science fiction and myth with accurate, up-to-date lunar science and space technology."

—DAVID MORRISON, founding director,
NASA Lunar Science Institute

"Beautifully written: evocative, witty, and truly informative. I thought I knew a lot about the Moon but I nonetheless learned all sorts of fascinating new things, and thoroughly enjoyed myself doing so. Combining science and science fiction is not easy and Oliver Morton manages it seamlessly and brilliantly! Superb."

—ADAM ROBERTS, author of *The Palgrave History of Science Fiction* and *The Thing Itself*

"A hymn to the Moon. I can't think of a wiser, more eloquent or better-informed companion for a journey around our natural satellite than Oliver Morton, whose poetic prose displays a breadth of knowledge not often found in science writing."

—Roger Highfield, director of external affairs, the Science Museum Group

"Our Moon, a spherical bit of unchanging inanimate rock, nonetheless captivates us with its romance and its beauty. Its tantalizing almost deceptive proximity makes it also a destination, both a past one and a promising one for the future. In this poetically written and informative book, Oliver Morton takes us through all aspects of this very familiar but very foreign territory, which has inspired stories and study for decades. What a remarkable achievement and one well worth exploring."

—LISA RANDALL, Frank B. Baird Professor of Science, Harvard University

THE
MOON

THE
MO
ON

A HISTORY FOR THE FUTURE

Oliver Morton

The Economist BOOKS

PUBLICAFFAIRS
NEW YORK

PublicAffairs
Hachette Book Group
1290 Avenue of the Americas, New York, NY 10104
www.publicaffairsbooks.com
@Public_Affairs

The Economist in Association with Profile Books Ltd. and PublicAffairs

Printed in the United States of America
Originally published in 2019 by Profile Books Ltd. in Great Britain.
First US Edition: June 2019
Published by PublicAffairs, an imprint of Perseus Books, LLC, a subsidiary of Hachette Book
Group, Inc. The PublicAffairs name and logo is a trademark of the Hachette Book Group.

The Library of Congress has cataloged the hardcover edition as follows:
Names: Morton, Oliver, author.
Title: The moon : a history for the future / Oliver Morton.
Description: New York : Public Affairs, [2019] | Includes bibliographical
  references and index.
Identifiers: LCCN 2019004627 (print) | LCCN 2019005523 (ebook) |
  ISBN 9781541768062 (ebook) | ISBN 9781541774322 (hardcover)
Subjects: LCSH: Moon—Popular works.
Classification: LCC QB581.9 (ebook) | LCC QB581.9 .M67 2019 (print) |
  DDC 523.3—dc23

LC record available at https://lccn.loc.gov/2019004627

ISBNs: 978-1-5417-7432-2 (hardcover), 978-1-5417-6806-2 (ebook)

LSC-C

10 9 8 7 6 5 4 3 2 1

To John Loft, my uncle,
John Hynes, my father-in-law,
and John Morton, my brother
beloved all

Her antiquity in preceding and surviving succeeding tellurian generations: her nocturnal predominance: her satellitic dependence: her luminary reflection: her constancy under all her phases, rising and setting by her appointed times, waxing and waning: the forced invariability of her aspect: her indeterminate response to inaffirmative interrogation: her potency over effluent and refluent waters: her power to enamour, to mortify, to invest with beauty, to render insane, to incite to and aid delinquency: the tranquil inscrutability of her visage: the terribility of her isolated dominant resplendent propinquity: her omens of tempest and of calm: the stimulation of her light, her motion and her presence: the admonition of her craters, her arid seas, her silence: her splendour, when visible: her attraction, when invisible.

—James Joyce, *Ulysses*

How many more times will you remember a certain afternoon of your childhood, some afternoon that's so deeply a part of your being that you can't even conceive of your life without it? Perhaps four or five times more. Perhaps not even. How many more times will you watch the full moon rise? Perhaps twenty. And yet it all seems limitless.

—Paul Bowles, *The Sheltering Sky*

Are we going to lunch? Or are we going to the Moon?

—Rip van Ronkel, Robert A. Heinlein
and James O'Hanlon, *Destination Moon*

CONTENTS

LIST OF ILLUSTRATIONS

THE NEARSIDE

INTRODUCTION

STRAWBERRY MOON

June 19th 2016, San Mateo County, California

THE CALIFORNIA SKY WAS WARM AND BLUE, ITS LIGHT STILL bright but softening. Shadows lengthened across dry grass towards San Francisco Bay as the train trundled south. In London, though, it was four in the morning, and it was in London that I had started my day. I was a third of a planet from home and I was tired.

I had come to Silicon Valley to talk to people about space and technology. In preparation, my head resting against the window of the carriage, I was reading a scientific paper on places where one might site a moonbase. I was not taking in the arguments all that well, but I was impressed by their breadth. The paper's Moon was mapped by laser, camera and radar, the shadows in its craters and sunlight on its peaks modelled by computers, its minerals assayed using electromagnetic radiation of every frequency—and neutrons, to boot. The data were as varied in source as in type; some came from Chandrayaan-1,

India's first lunar mission, launched in 2008; some from NASA's Lunar Reconnaissance Orbiter, which was launched the following year and had, six years later, sent its handlers a startling 630 terabytes of data. Some were older: from the Soviet Union's Lunokhod rovers, from America's Apollo landings, from the Lunar Orbiter missions that had paved the way for them.

From the range and weight of this material came pros and cons for various possible locations; a communications relay here is better than one there, this crater is more easily traversed than that one, the richer thorium deposits there do not make up for the more favourable solar power conditions here, and so on. The paper was not just making a case for this spot on the rim of Peary, a crater near the North Pole, versus that spot between Shackleton and Sverdrup, near the South. It was a performance—a demonstration to a world in general little interested in the Moon that, now all this detail was available, this was the sort of argument people could and should be having.

Then, out of the corner of my eye, I saw it, rising full.

I didn't catch the moment it broke the horizon; you very rarely do, unless you have planned accordingly. But it was still at the bottom of the sky, down where the logic of landscape requires the mind's eye to invest it with a size beyond that of its image as subtended on the retina. It looked as big as it looked distant, washed blue by the still-light sky, a depth as much as a brightness. You would never suspect that its spectral face was as stone-solid as the raised-up sea rocks of the California hills below.

It was, I realised later, a wonderfully apt place from which to see it. The train taking me from San Francisco airport to Mountain View was passing Menlo Park, where in the 1960s making maps of the Moon had been a rite of passage for the newly minted "astrogeologists" of the US Geological Survey. On Mount Hamilton, in the hills over which it was rising, is the Lick Observatory, where a pioneering photographic survey of the Moon was undertaken more than a century ago, and where those Menlo Park geologists would be sent, some eager and some unwilling, to inspect the object of their study.

Up ahead of me was NASA's Ames Research Center, the reason for my trip to Mountain View, home to the wind tunnels used to define the blunt re-entry-ready shape of the Apollo command modules, and home for a while to some of the rocks those modules brought back. Behind me, in San Francisco, was the home of Ambrose Bierce, author of one of America's great tales of the fantastic, "The Moonlit Road". Many gothic writers had used moonlight for unearthly effect before. In his story of three seemingly contradictory accounts, Bierce created a scene in which the flat, spectral light illuminated three truths, or none. A smooth light of inconsistencies; a single Moon of many stories.

The links were not all in the past. The little cluster of space-business start-ups outside the Ames centre had, until recently, housed Moon Express, a company which planned to launch the first commercial payload to the Moon. A few kilometres closer, on Bay View Boulevard, were the headquarters of Google, which was at the time the sponsor of a $30m set of prizes for landing a rover on the Moon which Moon Express, among others, was trying to win. On the other side of the tracks, in the hills above Stanford, was the home of Steve Jurvetson, a venture capitalist who had been an early backer of Elon Musk's SpaceX and nurtured his own plans for the Moon. It was at a meeting in that house that the moonbase-siting study I was reading had been conceived.

And beneath those hills, in the depths of the San Andreas Fault, the Pacific Plate and the North American Plate were responding to the full Moon's spring tide, just as they do every month. Tides do not, in general, trigger earthquakes, but they pull strongly and insistently enough for the supremely sensitive instruments of the seismologists to feel the Earth creaking gently at their touch.

But as the train took me down the valley towards the matte-silver-blue Moon, I thought about none of that. I was simply struck by how extraordinary it felt to be seeing the same object, at the same time, in two such strikingly different ways—to be surprised while reading the science by the beauty of the thing itself outside. It was not the feeling Walt Whitman expressed in "When I heard the learned astronomer", contrasting the drone of dry proofs and stale columns of figures with the silent, sublime power of the starlit night itself. It was its reverse: a deep sense that the different ways of seeing reinforced each other. A cognitive consonance of Moon as many stories, Moon as might be and Moon as always was, Moon longed for and Moon happened upon.

◗ ○ ◖

HAPPENED UPON. THE MOON, INCONSTANT IN APPEARANCE BUT constant in presence, is often seen but rarely looked for. Sometimes the wash of its light on buildings or landscape prompts you to seek it out in the sky, sometimes the clouds lit by its glow are hard to miss. More often, though, you just come across it unexpectedly, as I did on that train—or on the morning I am writing this, when the sliver of its waning crescent in the pre-dawn sky surprised me through my attic window. My decision to write this book has, unsurprisingly, made me more Moon aware, more likely to seek it out; I like to imagine that your reading it may do the same, at least for a while. But still I see the Moon as much by chance, I think, as by design, catching it from the corner of my eye.

This is appropriate. The Moon is essentially peripheral. It is rarely anybody's central concern, as a mountain or a sea might be, a person or a nation. It is off to one side, a lesser companion tagging along through the years. It differs from the rest of the yet further cosmos only by being close enough to see by day. In other ways it is as distant and as ineffable a part of the sky as the stars beyond.

But it is not just the nearest outpost of the elsewhere; it is also the furthest reach of here. It is in thrall to the Earth, its face cupped by the hands of gravity so strongly that it cannot turn its gaze away from ours. It is near enough for its pale fire to light the night, for its pull to raise the waters, to take the blame for madness. Its material is the same as that beneath our feet, and human feet have trodden it there as we do here. It defines the sky. It completes the Earth.

Far-off but been-there: unique but not inherently fascinating. Its story to date is in essence a simple one of achievement and abandonment. Its physical mysteries are few. The processes that formed its face are far simpler to understand than the chiselling and planing by which tectonic plates shuffle scraped-up sediments and submarine volcanoes into green hills above dry California valleys. Little of consequence has happened to it, and it may be that little ever will.

But something, at least, is going to happen soon. When I began this book there were five people alive who had walked the Moon. When it went to the printers, there were four. It is my firm belief they are handily outnumbered—perhaps by orders of magnitude—by the people now alive who will follow in their footsteps. The Return to the Moon is coming, and it will be undertaken by men and women from many more places, and with more agendas, than were in the American vanguard of 50 years ago. The space in my double vision on the southbound train, the space between engineering analysis trying so hard to become real and the real in the sky looking so determinedly unreal, will be filled. That space, the space between the Moon of the past and the Moons of the future, is the space of this book, a space of fact, speculation and digression, of ideals and inconsistencies, of the Moon itself, toe-stub certain as the rocks of the Diablo Hills, and of the Earthly ideas and concerns its peripheral light throws into sometimes irreconcilable relief.

Some fear or claim to fear or maybe even want to fear that a moon-based Moon would be a diminished Moon, a disenchanted Moon. They want the sky Moon only, not the rock Moon nor the scientific paper's Moon waiting for a base; they need those stories to contradict. Others want a Moon that is at one with their Earth, one at the far end of a journey longer than any the Earth can contain but which remains a journey like other journeys, thrilling, perhaps strange, but not, at heart, out of this world: the sort of place that has terminals for arrivals and terminals for departures.

The first, I fear, will be disappointed—but may yet find that the Moon can be re-enchanted if they feel for it in new ways. The others will find that the Moon is a very different thing to the Earth of contract and commerce. It is rock alone, and radiation, and that is all. It is utterly inanimate. That raises practical problems. It also raises issues beyond the physical. What is natural to a place that has no life? What is proper? What is wrong? Can such a place be a country or a home? Can it be a world of experience, or must it be always just a physical environment to be coped with technologically, always other, never cousin, never self?

● ○ ◐

THERE ARE TWO THINGS WE KNOW FOR SURE ABOUT GOING TO the Moon. It is doable. And it is undoable. Once you have gone, nothing necessarily detains you there or imposes any obligation to return. For the years of Apollo, the Moon carried life, and that was remarkable. But its subsequent lifelessness goes easily unremarked. It is of little concern to most of those who look up at it or forwards to the future; it matters almost not at all to geopolitics or the world economy or climate change. No one can argue convincingly that the Return will change that. The Moon may become important. It may not.

But if it is not necessarily important, it is never not lovely. Never the same, because the skies it sits in are never quite the same. And always the same, as well. Always the thing you saw when you saw it first, though that first sighting can never be remembered any more than you are likely to know for sure when you see it for the final time.

When your attention is snagged by a glimpse or a cue of the light—or, better, drawn by a human voice, standing near, exclaiming in wonder and asking you to share, as your mother did, that first certain but unremembered time—and you look on it again, how often does it disappoint? How often does its familiar strangeness not provide some small delight? How often does raising your eyes with those of another not tease out some soft sentiment? You may not look for long. But you seldom, if ever, regret paying that moment of attention to the little part of the world in the sky.

ITS PHASES

AT ITS FULLNESS IT IS A SIMPLE CIRCLE, EVERY PART OF ITS FACE lit directly by the Sun behind you. Any imperfections in its figure, such as a jitter to its edge, are due to atmospheric distortions, not its own irregularities: it is a near-perfect sphere. If it were reduced to the size of a billiard ball, the Moon would be as smooth as one.

At all other times the lit portion of its face is bound by two curves. One, the limb, is the edge of the Moon: on the inside of it, the Moon's surface; on the outside, the stars. It is always a semicircle. The other curve, the edge of the night, the line that brings an end to one lunar day and the beginning to the next, astronomers call the terminator.

After the moment of the Moon's fullness, this night-edge replaces the limb on the Moon's eastern side and starts to erode the disk. The Moon is now gibbous—less than full, but more than a semicircle—and waning.

The night-edge is less sharp than the limb, because the difference between Moon and not-Moon is sharp, but the difference between day and night is more complex. Summits can still be sunlit after the lowlands around them are in night; east-facing swales can enter shadow when it is still low-lit evening a stone's throw away. The line between lit and unlit is thus always a little incoherent.

On the Moon, it is also slow. At the equator the night-edge slides west at just 16 kilometres an hour; to the north and south, it is slower.

After two weeks the night-edge has eaten the sunlit part of the Moon away to almost nothing. A thin crescent arches between the polar horns where limb and night-edge touch, the six-to-nine-to-midnight rim of a clock. As it has waned, the Moon has drawn closer to the Sun. When full, it rises around dusk and sets around dawn, opposite the Sun in the sky. But the waning crescent Moon rises shortly before sunrise and sets in the afternoon. This Moon is of the day, not of the night.

Eventually, there comes a point when it sits so slim and slight in the Sun-soaked sky it can no longer be seen by the naked eye. Perhaps a day after that, the terminator meets the limb and there is nothing there at all. Every speck that is sunlit is invisible from Earth; every part that faces the Earth is in darkness. The Moon is new.

When the day-blind Moon returns, it trails the Sun down the evening sky reversed as in a poletopole mirror: its crescent runs from noon through three to six. The night-edge still crawls from east to west, but now it is chasing night away, leaving behind brightness as a squeegee leaves clean glass. The crescent fattens—waxes—until the night-edge stands straight, marking the first quarter. Then the Moon is gibbous again, slowly filling itself with borrowed light until, for a moment, it is whole again, all day again, a perfect circle.

Thus the phases of the Moon: full, waning gibbous, waning crescent, new, waxing crescent, waxing gibbous, full.

This regular cycle—from one full Moon to the next takes 29 days, 12 hours, 44 minutes and 3 seconds—has defined time since time was first defined. In the Islamic calendar a year is always 12 of these months, with each month beginning on the day when the waxing crescent is first seen in the evening sky. This means months can be either 29 or 30 days long. If the moment of the new Moon is early in the morning, then the crescent Moon may be seen the morning of the following day. If the moment of the new Moon comes late in the day, another whole day may pass with no visible Moon and the next month will begin the evening after. When the month thus drawn out is Ramadan, the month of fasting, that last day is long.

Months that are never more than 30 days long mean that the years of the Islamic calendar are shorter than solar years—the time it takes for the Earth to move round the Sun. In calendars where both the phase of the Moon and the season of the year are taken into account—lunisolar calendars like the Chinese and Hebrew ones—steps, such as introducing an intercalated month every few years, are taken to keep the lunar year and the solar year aligned. In Islam this is forbidden.

In solar calendars, such as the Gregorian calendar used in the West, the months and the phases of the Moon are not aligned at all. Nevertheless, there will be 12 or 13 full Moons a year, one in each month except, occasionally, February. There are various traditions for naming them. January's full Moon is the Wolf Moon, February's the Hunger Moon. March's Lenten Moon is sometimes called the Worm Moon or the Sap Moon. If the Lenten Moon falls after the spring equinox, then Easter is celebrated the following Sunday. If, as is more common, the first full Moon after the equinox is April's Egg Moon, then Easter comes the Sunday after that instead. May has its Hare Moon or Flower Moon; June, its Strawberry Moon.

Summer storms bring the Thunder Moon—which is also the Hay Moon, rising slow and low over evening meadows of grass kissed gold in the day. August brings the Grain Moon. The Harvest Moon is the full Moon closest to the autumnal equinox; it normally rises in September, sometimes in early October. After it come the Hunter's Moon, and then the Frost Moon, which if it rises particularly late in November or early in December becomes the Mourning Moon. The last Moon is the Cold Moon; then the Wolf returns.

Every few years or so, there are two full Moons in the same month. The second has recently come to be called a Blue Moon, whatever month it sits in.

Many names, many relations to the other signposts of the year. But always constant to its own rhythm: full, waning gibbous, waning crescent, new, waxing crescent, waxing gibbous, full. The Earth's skies offer no other regularity comparable.

- I -

REFLECTIONS

O N JUNE 24TH 2001, JUST AFTER THE SUMMER SOLSTICE, A telescope at the Observatoire de Haute-Provence in the South of France swung through the evening sky to fix itself on the waxing-crescent Moon sinking in the west. It may have been the first time that the telescope's mirror, 80cm across, had ever set out deliberately to gather moonlight. These days astronomers do not much care for the Moon.

In the 17th century its features fascinated them. Galileo Galilei's first telescopic observations of the Moon, and the deductions he drew from them, helped change astronomers' conception of what might lie beyond the Earth—and of what the Earth itself might be. Unlike the other bodies of the sky, the Moon had a physiognomy you could map, like an island, like a palm, like a face. And map it the telescopic Moon-watchers did.

By 1892 spectacular maps had been made, and great globes—or, rather, given that only one side is seen, half-globes—graced museums. The American geologist Grove Karl Gilbert, chief scientist of the US

Geological Survey, was able to announce, with some pride, that the Moon was in some ways perhaps better mapped than the continent on which he stood. None of it was as well known as the best-mapped bits of North America. But nor was any of it terra incognita, as some of central Canada or Alaska still was. All that was exposed to inspection—all of the Earth-facing nearside—could be recorded.

To a geologist like Gilbert, this panoply of strange landscapes formed by mysterious processes was a wonder; he owned himself "a little crazy on the subject of the Moon".* Many have followed in his craziness. Despite now having maps of far more distant, complex and dynamic bodies, some astrogeologists and planetary scientists remain in thrall to the faraway nearby of the Moon's surface. To astronomers, it was a bit of a bore. As the Victorian astronomer Richard Proctor put it, "The principal charm of astronomy, as indeed of all observational science, lies in the study of change,—of progress, development and decay. . . . In this relation the Moon has been a most disappointing object of astronomical observations." If astronomers had wanted to stare at unchanging rocks, they would have gone into geology themselves, or become stonemasons.

Worse than dull, it was actually damaging. When the Moon rides high at night, some of its bright borrowed sunlight is smeared across the sky, overwhelming the fainter, cosmically distant lights astronomers have come to prize. So by the 20th century almost all astronomers had come to shun the Moon, at least in their professional lives. They avoided the nights when it was at its brightest, shielding their instruments from its insistent stare as they waited for the dark of its departure. Only the amateurs kept Moonwatching, running their eyes and instruments over the raised relief of its features for the pure pleasure of its beauty and mystery, or in the persistent faith that at some point they would, indeed, see something change.

---

* A member of Congress unhappy about funding what he saw as the caprice of scientists noted that "So useless has the [Geological] Survey become that one of its most distinguished members has no better way to employ his time than to sit up all night gaping at the Moon."

● ○ ◐

IT WAS THUS VERY UNUSUAL TO FIND A PROFESSIONAL TELESCOPE
following the Moon as it slid down over the hills of Provence on that
evening in 2001. More unusually still, another fine telescope was
trained on the Moon just a few hours later, at Kitt Peak in Arizona. In
France it had set behind the Luberon hills, named for the wolves that
once lived in them, animals which were absent for most of the 20th
century but which have now, I believe, returned. In Arizona, with a
pleasingly symmetric regard for the role of nocturnal animals in lunar
folklore, it had risen over the granite of the Coyote Mountains.

The Moon's light is important to the lives of many creatures on
the Earth. The Peruvian apple cactus, for example, opens its enormous flowers only when the Moon is full. But as far as any researcher
can tell, neither wolves nor coyotes care all that much about it. They
howl on moonlit nights and moonless ones alike. Humans associate
howling with the Moon only because they associate the Moon with
the night, and howling with the night, and thus the howlers with the
Moon.

Coyote, according to a story that has been told in Arizona since
before it was Arizona, howls at the Moon because he used to be the
Moon. He took over that office from Crow, who the people had chosen before him but who had been too dark for the job. Crow had taken
the job on from Fox, the people's first choice, who had proved too
bright. The quality of Coyote's light proved superior to that of Crow
or Fox. But his presence in the sky proved troublesome. He would use
his vantage point to peek at women bathing, to reveal petty crimes, to
spoil games of chance. So the people called Coyote back down, as they
had Fox and Crow, and sent up another animal of similar colour: Rabbit. Rabbit curled up in the Moon, lay still, and didn't misbehave. He
has stayed there ever since. Look up and you can see him, body curled
up, ears drooping down.

Those visible features of the Moon, though, were not what the astronomers in France and Arizona were interested in. They were drawn

to the part of the Moon where the Sun was not shining, the part where features could scarcely be seen—the part that is lit only by earthshine.

Just as the Moon reflects the Sun's light to the Earth, so the Earth reflects the Sun's light to the Moon. And it does an impressive job of it. The Earth is larger than the Moon, and a better reflector; a full Earth, seen from the Moon, provides almost 50 times as much light as a full Moon does on Earth. And some of that earthlight, having travelled from the Earth to lighten the night of the Moon, bounces back whence it came.

It is when the Moon is a crescent that this light, sometimes called the ashen light, is best observed. At these times the Moon is between the Sun and the Earth; it is day on its Sun-facing farside, night on most of the Earth-facing nearside. That night is lit by the bright gibbous Earth. The earthlight means you can see the whole disk of the Moon quite plainly. The earthlit bit looks very dark—paradoxically it often looks slightly darker, to my eyes, than the surrounding sky. But it is clear that there is a whole Moon there, not just the bright shaving on one side. People sometimes call it the old Moon in the young Moon's sunlit arms.

The first person known to have understood the ashen light to be the light of the Earth was Leonardo da Vinci, in the early 16th century. Leonardo advised those who would be artists that "the mind of the painter must resemble a mirror"; the world was there to be reflected. And, sometimes, to reflect itself, as the Moon reflected the Sun to the Earth.

What was the reflecting surface? Choppy water, Leonardo thought. If the Moon were a true, smooth mirror, he pointed out, Earthly observers would see an image of the Sun glinting off just one point on its surface; he compared it to the highlight you might see as sunlight strikes a gilded ball decorating the gable of a high building. The fact that there is no such single highlight, he argued, must mean that the Moon was a set of mirrors all reflecting the Sun in slightly different directions—like a gilded mulberry (I love this image) or the waves of a fishing-boat-bobbing sea. Liquid seemed more likely than fruit. The lunar surface, Leonardo supposed, must be mostly sea—an idea that went with the flow of a long-running association between the Moon, raiser of tides and companion to rain clouds, and water.

The Earth, too, is largely covered by water. So it, too, must reflect the Sun. "If", Leonardo wrote, "you could stand where the moon is, the sun would look to you as if it were reflected from all the sea that it illuminates by day, and the land amid the water would appear just like the dark spots that are on the moon which, when looked at from our earth, appear to men the same as our earth would appear to any men who might dwell in the moon." The night-time Moon was lit by light reflected from the seas of the Earth just as the Earth's nights were lit by the seas of the Moon. It was an instance of the phenomenon of "secondary light" much discussed by artists in the Renaissance—the fact that there can be light when there was no visible source of it, as when light bounces off the wall of a sunlit room to illuminate a neighbouring room with no windows. Leonardo took the way artists thought about the painting of interiors and applied it on a scale larger than the Earth.

Galileo, who first brought the ashen light, and its explanation, to the attention of a broad public, was, like Leonardo, interested in the technical aspects of painting—indeed, he had taught a course in them. He also, though, had an interest that Leonardo lacked. He wanted to convince the public that the cosmos was not as they had thought it. And for this purpose the ashen light was particularly useful.

Most of the observations that Galileo wrote about in "Sidereus Nuncius" (1610; "The Starry Messenger"), a short but remarkably in-fluential book, were made possible by his then-brand-new telescope, with which he had started observing the Moon and other heavenly bodies the previous year. Since telescopes were rare, most of his readers had to take him at his word—and his illustrations, which show the artistic talent with which he had once wanted to make his name—as to what he saw. But seeing and understanding the ashen light required no such high technology. Galileo assured his readers they could easily see the effect for themselves if, when the crescent Moon was low in the sky, they so positioned themselves that the sunlit sliver was hidden by a chimney or wall. Once thus seen, the ashen light was easily, even naturally, understood as light that had bounced first off the Earth and then off the Moon, like sunlight reaching an inner room.

For most of his readers, this must have been a strange new thought. For Michael Maestlin, an astronomer at the University of Tübingen, and his pupil Johannes Kepler, then the court astronomer to Holy Roman Emperor Rudolph II, this part of "The Starry Messenger" came as no surprise. They had come to the same conclusions about how the Moon was lit with no need of telescopes. So had Paolo Sarpi, a priest and statesman in Venice whom Galileo knew and with whom he may have discussed the matter. It is no coincidence that these men were, like Galileo, members of the small band of Moonwatchers which took seriously the idea that Nicolaus Copernicus, a Polish cleric, had published more than half a century before: that the Earth orbited the Sun.

That belief is not directly related to understanding the ashen glow as earthlight. The ability of the Earth to reflect sunlight on to the Moon is independent of who is going around whom; it just requires that sometimes the Earth and Sun be on opposite sides of the Moon. A contemporary of Kepler's and Galileo's who believed the Earth to be the centre of everything could have embraced the same explanation for the nearly-not-light the crescent Moon held in its arms. But as far as is known, none of those Earth-centred contemporaries did; it was something only the Copernicans took note of.

Why did this way of understanding the Moon to be lit by earthshine fit with one conception of the universe, but not the other? Because understanding the Moon and the Earth as having the same powers of reflection required you to see the Moon and the Earth as the same sort of thing. That was part and parcel of being Copernican: the Moon and the Earth were planets like the other planets that orbited the Sun. It was a high conceptual hurdle for everyone else. The mediaeval world had followed Aristotle in believing that the Earth was of a fundamentally different substance from the Moon, or any of the other bodies that orbited it. The Earth was made of dull matter;

they were made of crystal, or fire, or some other rarefied substance. The Earth changed, but they did not. They moved, but the Earth did not.

Seeing the Moon lit by the Earth in just the way that the Earth is lit by the Moon went against that understanding. In Galileo's words, it drew the Earth "into the dance of stars". That choreographed companionship was at least as much a part of the Copernican revolution as the details of who orbits whom. The Earth became a planet—in the original sense of a star that wanders—and the planets became Earths, bodies as real as the world around you. Indeed, they might well be inhabited by people who saw them as worlds and for whom the Earth was a distant moving dot. They almost had to be: what would be the point of God creating uninhabited worlds? As the historian of art and science Eileen Reeves writes, there came to be "a nearly axiomatic connection, at least in the popular mind, between the theory of secondary light, the Copernican worldview and a belief in extraterrestrial life."

◐ ○ ◐

THE QUESTION OF EXTRATERRESTRIAL LIFE HAS DANCED WITH the science of astronomy ever since, as the Earth dances with the Moon and Sun: sometimes the ideas have been in opposition, sometimes in alignment. In the past 20 years they have sat in a striking conjunction. A great deal of astronomy is now justified to the public which pays for it as a search for life elsewhere.

And that is why, having ignored the Moon for decades, astronomers in Provence and Arizona found themselves peering so intently at its reflected earthlight at the beginning of this century. They were looking at it to learn what the signs of life on an Earth would look like if seen from afar.

In 1995, after decades of false alarms, astronomers started discovering planets around other stars. The light from such "exoplanets" was so faint that these bodies could be detected only indirectly—by their shadows as they moved across the face of their stars or by tiny shifts they caused in the spectrum of the stars' light. But those interested in life in the universe—astrobiologists, as they were becoming known—believed that, in time, bigger, better telescopes would let them see some of those exoplanets directly. And when that happened they would be able to look for signs of life.

Light from an exoplanet is light from a distant star that has gone into its orbiting exoplanet's atmosphere, been reflected or refracted back out into space and travelled on to the Earth. The many years of the last leg of that journey make no difference to the light; the fraction of a second in the exoplanet's atmosphere and bouncing off its clouds or surface leaves its mark. The molecules of the exoplanet's atmosphere absorb some wavelengths more than others. If astronomers could spread an exoplanet's light out, wavelength by wavelength, in one of their spectrographs, like a dealer spreading a pack of cards across green baize, they could pick out such effects; some cards would be missing, because some wavelengths had been absorbed in the exoplanet's atmosphere.

How might atmospheric chemistry offer evidence of life? Consider the atmosphere of the Earth and the other planets nearby. On Mars and Venus what atmospheric chemistry goes on is purely driven by sunlight; nothing on the surface is putting gases into the atmosphere that will react with each other. On the Earth life is endlessly churning out new gases, and the atmosphere is full of things that react with each other, such as methane and ammonia, carbon monoxide and oxygen and so on. In the 1960s James Lovelock, a British scientist and inventor, argued that this was a fundamental part of what it was to be a living planet. Life anything like the Earth's would have to use its planet's atmosphere both as a source of raw material and as a place to dump waste. The stuff it took out would not be the same as the stuff it put back, because it would only take out what it needed to use, and when it returned it that use would have changed it. Life will thus keep a planet's atmosphere from settling into the sort of equilibrium seen on lifeless worlds. The methane and ammonia and oxygen in the Earth's atmosphere are evidence of the flow of matter through a biosphere which uses the Sun's energy to transform that matter—of the biogeochemical cycles that knit the animate and inanimate into a living world. The insight that life kept atmospheres off-kilter in this way was an early step on the road to Lovelock's later Gaia hypothesis: the idea that life, through the creation of such disequilibrium, plays a fundamental role in keeping planets habitable, rather as riding keeps bicycles balanceable.

Not all of James Lovelock's ideas about Gaia have been widely accepted. The idea that life creates chemical disequilibrium in planetary atmospheres, though, has. By the early 21st century, theorists saw it as their best chance of diagnosing life over astronomical distances. But no one knew whether such observations might work in a practical setting. There is, after all, only one planet known to be life-bearing available for study—the Earth—and there are no observatories beyond the Earth that are set up to analyse earthlight spectroscopically.

Hence the observations made by astronomers in Provence and Arizona on a near-solstice night in 2001. The closest thing the sky has to offer to the experience of looking at another living world is to look at this one, reflected back from the distant mirror of the night-dark Moon.

● ○ ●

THE BELIEF THAT THE MIRROR MOON REFLECTS NOT JUST SUNlight but an image of the lands of Earth and the great ocean held to surround them dates back to ancient Greece, where it was embraced by some of the followers of Pythagoras. The arguments that this is clearly not the case go back almost as far. In "On the Appearance of the Face in the Orb of the Moon", the first treatise on the Moon—and one that would be read for its insights for more than a thousand years—Plutarch, a first-century-CE Platonist, argued strongly that the features seen on the Moon were those of the Moon itself, not reflections of Earthly geography. Features taken to be seas on the Moon did not share the shape of the great continent-girdling ocean of the Earth. What's more the Moon, unlike the image in a mirror, looked the same from every angle.

Nevertheless, the idea of the Moon as a reflection of the Earth persisted. In the early 17th century Kepler's patron, Rudolf II, apparently held it to be true, not least because he believed that he could see on the Moon the shapes of Italy, Sicily and Sardinia. Almost two centuries later Alexander von Humboldt recorded that the view was still held by educated Persians: "It is a map of the Earth . . . what we see on the Moon is ourselves."

There is no map; but what people see when they look at the Moon is indeed, for the most part, a reflection of themselves—of their preoccupations and theories, their dreams and fears. It has been used for such reflection, or projection, in science and fiction alike. The history of the Moon is a history of ideas about the Moon; and it is from those ideas that its future will grow. The Moon, always marginal, is hard put to support meanings of its own. It is there to be filled with the concerns of the big bright world that shines down from its jet-black sky.

In the second half of the 20th century, as the world was gripped by the prospect of wars that technology would make more dreadful than any before, the Moon reflected back conflict and contest; it was reimagined both as a battlefield and as the prize a winner might claim after a race. But those decades of conflict and contest also made it a more literal object of reflection.

Just before noon on January 10th 1946, a three-kilowatt radar transmitter which had been used for long-range detection of enemy aircraft sent a pulse of radio waves from Fort Monmouth, in New Jersey, to the rising Moon. Two and a half seconds later—the time it takes light, or radio waves in this case, to travel the 380,000km there and back—the signal returned. The engineers had become, at least in their own eyes, the first humans to touch the Moon.

At a time when the brilliant light of the atomic bomb threw a new shadow across the future, this seemed, to the cognoscenti, a wonderful event. As the trade magazine *Radio News* put it in an ecstatic editorial:

Radar has now taken us out of this world, plunged us into the infinite, challenged the universe with spears of radio impulses that have prodded the moon and returned to open new floodgates of human mental activity. No longer can the defeatists tell us that mankind must settle down to the uninspiring prospect of making the most of our little world . . . [e]ven as radio has been a vital factor in shrinking our own world, so does radio now break our fetters and carry us to worlds beyond.

True to its marginal nature, though, the Moon itself was somewhat incidental to what the Army called "Project Diana". Radio operators depended on the ionosphere, a layer of charged particles in the Earth's upper atmosphere which bends and reflects radio waves, to get signals to travel over long distances. A better understanding of the ionosphere thus had great practical value, and getting radio waves to go all the way through it and then bounce back promised to improve that understanding. What was more, if space travel were to become possible—as the advent of long-distance rockets and nuclear power were leading some to suspect that it would—it would be important to know that the travellers could stay in touch with the planet they had left behind.

Indeed, it was possible that radio communication might not just be necessary for the support of space travel: it might be its purpose. Shortly before Project Diana, Arthur C. Clarke, a young British electrical engineer who had worked on radar during the war, wrote a paper on the role that "extraterrestrial relays"—communication satellites, in particular those in "geosynchronous" orbits that take 24 hours to circle the Earth, and thus seem fixed at a specific location in the sky—could play in providing worldwide radio and television coverage. "We have as yet no direct evidence of radio waves passing between the surface of the earth and outer space," he noted, "[but] given sufficient transmitting power, we might obtain the necessary evidence by exploring for echoes from the moon." I do not know that the men of Project Diana knew of Clarke's seminal paper, but their counterparts in the US Navy did, and so did some in the press. On February 3rd 1946 the *Los Angeles Times* ran a front-page story on the idea, noting of Clarke's proposed Moonbounce test that "the US Army Signal Corps has just done this."

Project Diana thus showed both the feasibility of communication satellites and that the Moon could function in such a role. The first was the more important. Some subsequent military projects used signals bounced off the Moon instead of off the ionosphere when keeping in touch with far-flung components of the Cold War apparatus. But once Clarke's communication satellites were realised, they usurped the Earth's natural satellite.

Not all the radio reflections from the Moon were deliberate. In 1960 there was consternation when the displays at an American early-warning radar in Greenland suddenly lit up with unexpected returns: the Moon was rising slap in the middle of the radar's beam and bouncing its signals back to them. Contrary to some accounts, this did not look enough like a missile strike to cause a genuine false alarm. But it did lead the Air Force to reprogram its computers to ignore any radar return with a delay of more than two seconds, thus ensuring that the Moon could not confuse any further operations.

Scientists, for their part, made use of such reflections to further their understanding of the Moon's surface. But not all the subsequent radar work was scientific. In the 1960s the Soviet Union deliberately aimed beams from its newest and most powerful missile- and satellite-tracking radars at the Moon, presumably to calibrate them, sometimes for half an hour at a time. This offered the United States a nice opportunity for some celestial espionage. William Perry, an electrical engineer who would later become US secretary of defence, led a secret programme that studied the Soviet radar by using a radio-astronomy dish at Stanford to pick up its signals scattered back from the Moon. Filtering out signals from a local taxi company's dispatchers, who were using the same wavelength, was a headache, but what they managed to discover about the radar's capabilities showed that it was not sophisticated enough to target anti-ballistic-missile defences.

As far as I know, spooks no longer use the Moon this way. Just as artificial satellites provide better communication channels than our natural one does, so they are probably better suited to the needs of such snooping. Radar beams are still occasionally bounced off the Moon for scientific purposes. Other radiation is sent there and back, too: the Apollo missions left little mirrors behind, and various observatories regularly bounce laser beams off them to measure precisely how far away it is and how fast it is receding.

And if satellites have robbed the Moon of a professional position in the radio-reflecting business, it still competes on an amateur basis. Radio hams have no way of getting a signal farther than by bouncing it off the Moon and back to the Earth, and since some of them measure their prowess by the distances over which they communicate, excellence in EME (Earth-Moon-Earth) communications, which require big antennae, good equipment and a lot of patience, is a badge of pride in parts of the community.[*]

Artists use it too. In the 1980s Pauline Oliveros, an avant-garde composer and musician, staged an event called "Echoes from the Moon" at various sites. She would send sounds made on stage down a telephone line to a ham radio enthusiast, who would send them to the Moon, pick up their reflections and play them back. The sound of a gas-pipe whistle and Tibetan cymbals came out particularly well, she believed, after the first experiments; at later shows she played her accordion via the Moon. On some occasions audiences got to send their own voices there and back, too (at one such show this was achieved using the same Stanford dish that Bill Perry had used to snoop on the Russians). The audiences loved it.

In 2007 another artist, Katie Paterson, translated the notes of the first movement of Beethoven's "Sonata 14 in C Minor"—the Moonlight—into Morse code. She had the resultant dots and dashes bounced off the Moon and translated the signal that was reflected back into a score that could be read by an automated piano. The resultant installation, "E.M.E.", is magnificent. Various science fiction writers

---

[*] Bouncing lasers off the Apollo retro-reflectors, though, is not a plausible pastime for amateurs; the episode of "The Big Bang Theory" called "The Lunar Excitation" is misleading in this respect.

had previously imagined the Moonlight Sonata being played on the Moon. But none had imagined it being played via the Moon—and being reinterpreted through the imperfections of that lunar reflection. Some notes are lost, some are changed. The stately progression and even tempo of the music highlight the holes left by notes lost in transmission; the brokenness personalises the otherwise perfect technology of the playerless piano. The surface of the Moon, as touched by technology, is made present as a set of random absences. In all reflections there is loss.

● ○ ◗

THE ASTRONAUTS OF APOLLO 8 TOOK NO MUSIC WITH THEM, NOR did they bring any back. Cassette players weren't allowed on spacecraft until the following year.* And they didn't actually touch the Moon. But over Christmas 1968 the crew of *Apollo 8*, Frank Borman, Jim Lovell and Bill Anders, became the first men to follow in the path of earthlight and Project Diana and travel to the Moon and back, encased in the honeycombed aluminium and steel of their command module, fed packaged food and bottled air, recorded on magnetic tape, pushed down by thrust, floating free, sometimes spacesick, sometimes sleepless, endlessly following instructions, doing their jobs, isolated, crowded, keeping it together, experiencing it. Seeing. Changing.

The Apollo programme's later command modules had names of their own: *Gumdrop, Charlie Brown, Columbia, Yankee Clipper, Odyssey, Kitty Hawk, Endeavour, Casper* and, finally, *America*. The *Apollo 8* spacecraft had no name other than that of the mission itself. It took off from Cape Kennedy at 07:49, local time, on December 21st. The

---

* According to Brian Eno, whose album "Apollo", the soundtrack to the film "For All Mankind", is a sublime meditation on the lunar experience, all but one of the astronauts who took tapes to the Moon took some country and western music; it is part of the reason why his piece makes use of steel guitar.

engines of its Saturn V booster lifted it to orbit in less than 12 minutes.
The crew started to check out the systems on their spacecraft

—OK, this is a tape recorder test: 1, 2, 3, 4,
5, 6, 7, 8, 9, 10; 9, 8, 7, 6, 5, 4, 3, 2, 1.

as the Earth passed beneath them at just under eight kilometres a second, blue and green and white and big.

—I mean, let's get comfortable.
This is going to be a long trip.

Three hours later

—You got the redundant components
checked, Bill?

the third stage of the Saturn V, to which the command module was
still attached, fired up again

—Three, two, light ON. IGNITION

and flung the spacecraft away from Earth. Three days later it crossed
the Moon's orbit just ahead of the Moon itself, a mouse scurrying over
railway tracks in front of an express. As the Moon barrelled past a
few hundred kilometres behind them, its bulk cut the astronauts off
from Mission Control. They prepared the spacecraft's engine for the
four-minute burn

—Jesus, four minutes?

needed to put them into orbit around the Moon. Helium forced fuel
and oxidizer out of the propellant tanks and into the engine. There
was no order sent up from Earth to tell them when ignition should
be. There was, as far as they were concerned, no Earth. Just empty
space, and a single engine stopping them from falling even farther
out into it.

—Longest four minutes I ever spent.

Until this point they had not seen the Moon. The spacecraft had been pointed the other way. Once in orbit, though, they turned their vessel around and saw a wall of darkness rimmed by a curve of stars. They circled the blackness for ten minutes before

—I'm just going to move my eye away, because the sun's going to peek over here any second now.

the horizon in front of them overtook the night-edge, and light returned to the world.

For three orbits they stared down as the Moon passed under their windows at two kilometres a second

—It looks like a big—it looks like a big beach down there.

disturbingly real but oddly indistinct. In the 40-minute periods they spent over the daylit surface they tried—and, frequently, failed—to pick out the features

—Hey, you know something; it's gray, huh?

of the strangely lit sideways-seen land below—features which, a few months later, would guide their comrades on the *Eagle*, Apollo 11's lunar module, down to the first landing. In the 40-minute nights they attended to their own needs, and those of their vessel.

It was only on the fourth orbit, when the spacecraft's orientation had changed, that they had the vision for which their mission will always be remembered. Shortly after they had swept from night to day, a bright, colourful complexity came past the limb of the Moon and into view

—Oh my God! Look at that picture over there! Here's the Earth coming up. Wow, that is pretty.

and life returned to the world.

The men scrambled to get a photograph, tourist-happy and some-thing more. Borman claimed to have caught it first, in black and white, its terminator just above the limb of the Moon. Anders

> —Hand me that roll of color quick,
> would you?

caught it higher, a bit more than its own width clear of the limb, its white and blue and green and ochre a comment on the grey waste be-low it, a contradiction.

It has been called

> —Oh that's a beautiful shot.

the most important picture of the 20th century. If you have eyes and live in a world of books or screens, I know that you have seen that picture almost as surely as I know that you have seen the Moon in the sky.

◐ ○ ◐

TO REFLECT, LITERALLY, IS TO BEND BACK. THE IMAGE ON THAT roll of colour turned something which had been thought important because of where it was going into something which was important because of where it had come from. An unparalleled achievement be-came a passing platform from which to appreciate the true prize: the place left behind.

When originally presented to the press, the picture showed the lunar surface as a wall on the right, the Earth in darkness to the left. It is an orientation that makes them both properly Copernican bodies in space—the same orientation that George Lucas used when the Death Star came into view around the planet Yavin in "Star Wars" nine years later. It was evocative in part because it seemed so strange.

By the time the picture was on the cover of *Time* and prompting reflections by writers of all sorts, it had been rotated 90°. The Moon was a landscape with a horizon, the blackness of space was the sky and the Earth was rising into it. This is the orientation Stanley Kubrick had used for similar shots in "2001: A Space Odyssey" the year before, one that feels both more cosmic and more personal than the side-to-side arrangement. By evoking the familiar rising Sun, or Moon, it locates the viewer; making sense of the picture this way puts you into it.* Indeed, in a rather pre-Copernican way, it makes you its centre.

"Earthrise", thus displayed, has three elements. There is the background black, achromatic and featureless; you do not need to see it as an echo of Kazimir Malevich's iconic "Black Square" (1915), which the artist proclaimed to be the first picture unrelated to any real-world subject, to see it as an unsurpassable negation.

Apparently superimposed on the picture's lower half, a bright horizontal field, textured more than featured. In 1969, well after the imagery of "Earthrise" entered the culture, Mark Rothko used the same bisected construction, black above, textured grey below, in one of his last untitled paintings. It was a painting, he said, about death. Is it in the grey, the black or their juxtaposition that death is to be found? He did not say.

The third element of "Earthrise" is within its black upper half, a second brightness. That light in the darkness is the Earth waning gibbous, slightly more than half of its face in the Sun, its night-edge a relaxed bow over the limb of the grey-lost Moon. This is not death: this is not nothing; this is life, proud and sharp.

Just above the night-edge, almost central to the disk (and thus more or less the closest bit of the Earth to the camera) sits Ascension

---

* The different resonances of the orientations come in part from the syntax of cinema. Drama tends to move horizontally—movement across the screen takes you forward in time. The left-to-right reveal of the Death Star (a space station easily confused with a moon) as it moves out from behind the planet Yavin is inherently plot-y: Lucas uses it to build narrative tension. Moving the camera vertically signifies a pause, or a timelessness—a looking up, a heightening of tone. Thus Kubrick's alignments of Earth, Moon and Sun seem to transcend plot or drama, taking on the more sublime sense of a cosmos defined only by vastness and viewpoint.

Island, the invisible-at-such-distance speck of volcanic rock where, at the very moment the picture was taken, the antenna at a tracking station known as the Devil's Ashpit was monitoring *Apollo 8*'s radio transmissions.

People had imagined such sights before. But when artists had drawn them, they had almost all got them revealingly wrong. The Earth seen from the Moon in such prefigurings was almost always a school-room globe, dominated by the outlines of familiar continents: the world as mapped by humans and known by humans. Seen in itself, though, it was not a world but a planet, strange and changing, its features barely recognisable, its character unmistakeable. Not a representation, but a presence.

In "Earthrise" the North Pole is to the right of the disk, below the night-edge, invisible in its midwinter solstice. The South Pole blazes midsummer bright on the limb's upper left. Indeed, all the limb shines white. The cloud which sits over the seas and hides the coast of Brazil is as bright as the ice of the Antarctic. The contrast with the blackness beyond is absolute.

Within that unbroken white border, the disk's most obvious features are its curving weather fronts. They wrap themselves clockwise over the Southern Ocean, widdershins over the Atlantic to the north, their ceaseless movement expressed in the straining of their curves. More permanent features are harder to make out. To the left, just above the night-edge, is the sunlit coast of the Namib Desert, though you would be hard put to spot it without being told what to look for; to the right, more prominent, are the bright sands of the western Sahara. The only clearly identifiable geographical feature is a discrete inflection of sharply seen North African coast called Ras Nouadhibou. In 1441—at the beginning of the European voyages of exploration which the Apollo missions sought, in their way, to emulate—Nuno Tristão, a mariner from the court of Henry the Navigator, became the first of his countrymen to sail past that cape. On the same voyage he became the first European to take slaves from the shores of West Africa. Today the bay behind the cape is a graveyard for abandoned ships.*

The land is darker than the cloud, the ocean, contrary to Leonardo's expectation, darker still—except for one central spot, a bit more than half of the distance between the night-edge and the limb, a spot which shines in a manner all its own. It is the part of the South Atlantic in which, at this particular moment, for this particular geometry, the sea's surface really has become the mulberry mirror Leonardo believed it to be; the afternoon sunlight is hitting it at just the right angle to bounce straight towards the Moon rising in the eastern sky. This wave-bounced highlight, known in optics as a specular (meaning mirrorlike) reflection, has a different quality to that of the cloud tops. It is burnished: bright like metal, not like snow.

But it is all the same light. The men of *Apollo 8* were too far away to see any human lights from the cities of Africa, which would then have been the dimmest of embers; all of Nigeria, at that time, consumed less electricity than a small city in America. They saw no fires, no volcanoes, no lightning. Everything they saw, on Earth and Moon alike, they saw by sunlight.

Yet how different the two equally sunlit bodies look in that picture: one complex and fine-featured, dynamic and contained, its stillness purely a function of the shutter, like a snapshot of a dancer in mid-air; the other a partial expanse of greys that might go on forever, uneven but subdued, slightly lacquered-looking, still. It seems unfinished yet long abandoned, in need of something but incapable of anything.

The Moon reflects just 12% of the light it receives from the Sun. The rest it absorbs, like tarmac in a desert. The rocks close to its sur-

---

* It is also, in an odd coincidence, one of the best places in the world to buy a bit of moonrock. None of the rocks brought back by later Apollo missions are, officially, in private hands. But about 1 in 1,000 of the meteorites that land on Earth comes, originally, from the Moon, and in general meteorites are most easily found in deserts: the dryness stops them from becoming weathered, and thus less distinctive; the open landscape allows them to stand out; and desert dwellers tend to be sharp-eyed and attentive. The markets of Nouadhibou are among the best places to buy meteorites, including lunar ones, gathered by Saharan nomads.

face heat up to more than 100°C during their 340-hour day. In the 340-hour night they cool as the energy the Sun imparted to them slips back into the cosmos as heat; by late in the night their temperature is down to minus 150°C. But this huge thermal swing achieves almost nothing. The Sun's heat penetrates only a metre or so down into the dust, rubble and rock; lower than that, nothing changes. Energy comes down from the sky, energy goes back out to the sky, energy makes no difference worth the telling in between.

The Earth absorbs 20% less energy from the Sun, per square metre, than the Moon does. But it is able to do infinitely more with it.

The flow of energy which, on the Moon, just heats up a thin skin of rock, drives perpetual change on the surface of the Earth. Every second, 16m tonnes of water evaporated by the Sun at the surface rise into the sky. In the cooler heights that water vapour condenses into clouds in smooth, soft layers, clouds in tall, stormy towers, clouds like hawks and handsaws and whales, into tiny clouddrops and fattened raindrops, hard-falling hail, high-floating ice and every other lightness and darkness, softness and hardness of the sky. The condensing of the water vapour releases to the atmosphere the energy that evaporating the water took from the sunlit surface, fashioning gradients of temperature and pressure that all but ceaselessly stir the air.

The oceans, too, move heat, shipping it in bulk from the tropics to the poles, the journey twisted by the same Coriolis force as those curls of "Earthrise" cloud which go one way in the north, the other in the south. These polewards flows redistribute about 5% of the energy the Sun delivers to the Earth, the heat of their waters, the play of their currents and the Sun-driven stirring of the air above forming storms and winds and monstrous waves—as well as moments of sudden sharp stillness, nights of clarity as brittle as ice and fogs that settle motionless for days. And all this happens simply because it can. Simply because the presence of an ocean and an atmosphere allows it to—in fact, requires it to.

And there is more. Life has furnished the Earth with leaves as great in their aggregate surface area as the continents themselves. Along with the less familiar, but just as vital, photosynthetic membranes of algae and bacteria, those leaves take up about a thousandth of the incoming sunlight and use it to turn some of the air's carbon dioxide and some of the water that rises and falls from the skies into oxygen and biomass.

This is the transformation that keeps the atmosphere off-kilter; this is the transformation that is basic to the Earth as a living planet. Almost everything on Earth that lives does so thanks to that transformation; all the energy that is ever taken from eating another living thing comes ultimately from sunlight. Every twitch of a muscle and every spark of a nerve is sunlight, too.

Few who look at "Earthrise" appreciate these climatic, oceanic and biogeochemical dynamics in any detail. But almost all appreciate what they mean: the orb they are seeing varies and is various, it is changing, dynamic, a living world above the softly bleak unworld below. The twofold message of "Earthrise" is simple: that the Earth is there, in the sky, alive; and that someone alive, there in the sky, has seen it.

● ○ ◑

THE EARTHLIGHT OBSERVATIONS MADE AT THOSE OBSERVATORIES in Provence and Arizona in the early 2000s were a recapitulation of "Earthrise". Using what returns from the Moon to get another new view of life on Earth, they and subsequent studies have captured the same sense of the planet's off-balance dynamism, though in numbers instead of pictures.

Here are the features of the Earth the ashen light of the Moon can reveal. The chemical disequilibrium that James Lovelock first appreciated as a sign of life can be seen in the spectrum of the atmosphere; the presence of oxygen and methane, gases eager to react with one another, requires there to be continuous supplies of both—supplies we know to be driven by life. The presence of oceans can be inferred from glints like those the crew of *Apollo 8* saw in the South Atlantic—such specular reflections polarise light, and the polarisation is measurable at a distance.

It is not just the presence of seas and life that the earthlight reveals. Because oceans are dark and continents light, the distribution of land and sea can be worked out from regular changes in brightness. Obviously, the regular 24-hour period of those fluctuations in itself reveals the length of the day; more subtly, the copious noise in those fluctuations provides a sense of how much cloud cover the planet enjoys and how much that cover varies over time. Indeed, it was to measure the Earth's total amount of cloud cover, and to look for long-term changes in it as a result of global warming, that scientists started in the 1990s to systematically measure the earthlight from the Moon.

The presence of plants can be seen, too, through an intriguing feature called the "red edge". The pigments in plants absorb almost all the visible wavelengths of light to power the great transformation of photosynthesis. The only visible wavelengths most of them do without are the green ones, which they reflect; that is why leaves look green. Beyond the visible wavelengths, though, they exhibit another colour. Leaves reflect infrared radiation, which has wavelengths a bit longer than those of visible light, very efficiently indeed. This is not a matter of happenstance; it is an evolutionary necessity. If leaves absorbed all the infrared energy that hit them, as well as most of the visible light, they would get too hot. As a result, when we look down on forest canopies from the mountaintops they are dark to our eyes, but the right camera sees them as dazzling bright with cast-off infrared.

Leaves are common enough on Earth's surface that the light of the Earth as a whole shows this effect. As you go from detecting the red wavelengths of light to the longer infrared ones, the planet brightens sharply: the spectrum has a "red edge". Such a feature, it is argued, could not be explained by mere minerals. Only a surface that has evolved to make optimal use of some wavelengths of light while prudently rejecting others could provide such a threshold. Seeing such an edge in the reflected starlight of an exoplanet would be an indicator of similar evolution there, as well.

Thus earthlight has taught astronomers a lot about what is now widely taken to be the greatest near-term challenge of their trade: how to find evidence that distant exoplanets are, like the Earth, alive. By using the passive, lifeless mirror provided by a body that, for the most part, they scorn, they have found out what can be learned about the Earth's true doubles, light years away.

The irony goes deeper than that. The appreciation of the dim earthshine on the Moon by Galileo, Kepler and their contemporaries was a key part of the Copernican revolution—part of the discovery that the Earth was not the centre of the universe, that it was not so special, that it was a planet among many orbiting a star among many. Ever since, astronomy has had something of the humblebrag about it: look how skilfully and powerfully, the astronomers boast, we can show just how not-special we are. As their science spread its horizons wider, studying whole galaxies and clusters of galaxies, reaching back to the Big Bang itself, its observations were studiously framed so as to drive home the cosmically diminished significance of the species doing the observing.

But since, and possibly because of, "Earthrise", it has been possible to observe this trend reversing. The Earth has not been restored to the centre of the astronomers' universe. But a sense that what is special about the Earth—life—is somehow central to the universe, too, has grown. That, as with Apollo 8, looking out matters most as a way of looking back. That it is the point of departure, not the destination, that matters.

Astronomy is thus now increasingly treated, especially by its popularisers, not as a way of understanding the vast and inhuman universe but as a way of understanding, through that universe, humans and the Earth. One of the ways this is done is by concentrating on origins. That the origin of the universe is "where we come from" is treated not as a truism—where else would we come from?—but as a genuine addition to humankind's self-knowledge, thus rendering the search for further information about the quantum fluctuations of the Big Bang oddly personal. Rather than revealing human insignificance, such studies of the ineffably vast and ancient are seen as, or perhaps sold as, deepening our connection to the cosmos.

The same sense of connection stretches, as the prospectus for a new space telescope recently put it, "From Cosmic Birth to Living Earths". It is now a commonplace in lay accounts of astronomy that the search for other planets which carry life is a priority second only to understanding the origins of the universe—and perhaps surpassing it. Astronomy as encountered by the public (and, increasingly, as funded by them) is ever less about stars, ever more about exoplanets. The great thrust of the discipline is no longer to find things ever farther away that show the Earth to be ever less significant. It is to sift through the endless streams of starlight for something that looks like earthlight— to find something, somewhere in the galaxy, as special as the Earth. Something which might look back at us.

Just as the earthshine visible on the Moon had a walk-on part in the great decentring of the Earth, now it has a walk-on part in this anti-Copernican astrobiological recentring of the heavens on life and its Earth homes. It is a way in which the unworld Moon teaches the world of Earth what life is, and how to see it.

There may be other such lessons. To live there, or to imagine living there, and find how strange it really is to take life from the planet that has shaped it and which it has shaped in its turn. To step outside the endless flows through which that reciprocal shaping had been achieved, leaving them behind you in the sky, and to come to terms with a life purely technological, purely human, constantly kept from death only by thin, manufactured walls. To learn to what extent such a future can be a continuation of the history that has come before and to what extent must it be a break with it—or a dead end. To stand on a grey, foreshortened plain, your shadow cast, in an alien night, by the light of the world that gave you birth.

The Moon still has new reflections to offer.

ITS SIZE AND APPEARANCE

ITS MASS IS AROUND 73 MILLION TRILLION TONNES, 50 TIMES THE mass of the Earth's oceans, but just 1.2% of the mass of the Earth as a whole. If you were to slice straight through the Earth at 55°S—the parallel that passes through the southern tip of South America—the end-of-an-egg cap you would cut off, 840km deep and 7,000km across, would have roughly the same mass as the Moon.

This is less than the mass of any of the solar system's planets; it is a tenth the mass of Mars. Three of Jupiter's moons—Ganymede, Io and Callisto—are more massive than the Moon. So is Titan, Saturn's largest moon. But those moons represent only a tiny fraction of the mass of the mighty planets they orbit—about a five-thousandth—as opposed to the Moon's eightieth of the Earth's.

The Moon is more than five times the mass of the dwarf planet Pluto and some 25 times the mass of all the asteroids in the asteroid belt put together.

At least 95% of this mass is rock. A small amount of it forms a crust, which is about 40km thick, on average; the rest of it forms an underlying mantle. The Moon's iron core, if it has one, is less than a twentieth of the mass of all that rock. It is no more than 300km across and mostly, perhaps entirely, solid. Unlike the Earth's iron core, which represents 30% of the planet's mass and is mostly molten, whatever core the Moon may have produces nothing noticeable by way of a magnetic field.

The Moon's atmosphere would, if warmed up and pressurised to the conditions at the Earth's surface, barely fill a parish church. For almost all practical purposes, it has no atmosphere at all.

Its surface area is 37.9 million square kilometres, about a quarter of the area of the Earth's continents. That makes it smaller than Asia, a bit larger than Africa, and significantly larger than all the others. A straight tunnel between the Moon's poles would be 2,474km long—in terms of Africa, the same length as a tunnel from Cairo to Nairobi. A road around its equator would be 10,921km long—equivalent to driving from Cape Town to Addis Ababa and back. Imagine Africa stretched, snipped, spindled, stuffed and sewn back up to form a sphere around that Cairo-Nairobi axis and you have something of a sense of it.

About 17% of its surface is made up of dark, low-lying plains called *maria*: the plural of *mare*, which is Latin for "sea". Almost all these plains are on the Earth-facing nearside. The largest, in the west, is *Oceanus Procellarum*, the ocean of storms. It is half the size, more or less, of the Sahara. Above it lies *Mare Imbrium*, which covers about 2 million square kilometres, an area roughly the size of the Congo Basin.

To the east of *Imbrium*, names are less nautical and more allusive; here we find not rain and storms but serenity, tranquillity and fertility. Going east, *Mare Serenitatis*, roughly the size of Nigeria, gives directly into the slightly larger *Mare Tranquillitatis*, roughly the size of Chad. Two further, less circular maria branch off *Tranquillitatis: Mare Nectaris* and *Mare Fecunditatis*. To cultures that see a rabbit in the Moon, the recumbent *Y* of these seas is a pair of floppy ears.

Mare Crisium, the sea of crises, sits to the north of these other seas, a dark eye surrounded by brightness: as the sunlight spreads across the surface in the early days of the waxing Moon, *Mare Crisium* is the first feature to become distinctly visible; as the Moon wanes, it is the first to be lost. Luna City, the setting for Robert Heinlein's "The Moon Is a Harsh Mistress", the most influential novel ever to be set on the Moon, sprawls through tunnels beneath it.

On the far side of the Moon, the side that the Earth never sees, there are just two small seas: *Mare Muscoviense* and *Mare Ingenii*, the seas of Muscovy and cleverness. They cover an area slightly larger than that of the two islands of New Zealand.

The brighter parts of the Moon are called the highlands. Whereas the maria are not seas, just plains, the highlands are, for the most part, high. Where they meet the maria there are often mountain ranges, mostly named after counterparts on Earth; those around *Mare Imbrium* are *Montes Alpes*, *Montes Jura*, *Montes Carpatus*, *Montes Caucasus* and *Montes Apenninus*. The Apennines, 400km long, are perhaps the most impressive, their peaks rising as much as 5km above the non-rainy plains—similar in stature to the Rwenzori Mountains in East Africa, known in antiquity as the "Mountains of the Moon".

The most striking highland feature, when the Moon is full, is the crater Tycho, in the nearside's southern highlands. It is the nexus for a starburst of bright linear "rays" that spreads across half the hemisphere. There are also bright craters in some of the maria, such as Copernicus, south of *Mare Imbrium*. The brightest of all is Aristarchus, in *Oceanus Procellarum*.

Tycho is the youngest major feature, probably about 100m years old. It dates from a time when, on Earth, the rift which created the Atlantic had just started to pull South America out of what is now the Bight of Benin. Copernicus is about 800m years old, which makes it older than all the fossilized animals of the Earth. And Copernicus is still, in terms of lunar geology, a recent structure. Most of the maria are four times older than Copernicus. The highlands are more ancient still, well over four billion years old.

All but the very oldest rocks still extant on Earth are younger than all but the very youngest features on the Moon.

The great bright craters and dark maria discernible to the naked eye have not changed one bit over the time that there have been humans to look at them. Every person who was ever born who could see, and who lived long enough to focus their eyes and walk outside into the night, has seen the same Moon as all who look at it today. More humans have gazed in wonder at the surface of the Moon than at any other solid object in the universe.

- II -

THE FACE OF THE MOON

WHEN ALBERTUS MAGNUS, ONE OF THE GREAT ARISTOTE-lian schoolmen of the Middle Ages, looked at the patterns on the face of the Moon, he saw a beast with its head in the west and something on its back resembling a tree. In the east, there was a man who might have been leaning on that tree. Some saw that eastern homunculus as Cain, the first murderer, or as Judas Iscariot; others said he was just a peasant who had been banished from Earth for taking wood to which he was not entitled from some lord's lands.

It is possible that the dog and thornbush Shakespeare gave to Robin Starveling so that he could present the Moon in "A Midsummer Night's Dream" referred to the features Albertus saw as a beast—*Oceanus Procellarum* and attendant smaller maria—and the crown of a tree—*Mare Imbrium*. It is hard to tell, though. As far as is known, at that point no one in the Western world had actually made, let alone preserved and labelled, any drawings of the features they saw when they looked at the Moon, or given them settled names.

Today, in an age saturated with images, this seems remarkable—I certainly found it so when I first learned that it was the case. But no one remarked on it at the time. Symbolic representations of the crescent Moon—with a nose and a face added in profile if personification was in order—seem to have been entirely sufficient and were widespread, not least in coats of arms and the flags of Islam. Pictures of its features, as opposed to its form, were not. If you wanted to know what the spots on the Moon looked like, you could look at the Moon. Why record things all could see, which did not matter and which no one was in danger of confusing with anything else?

The first answer seems to have been this: because recording the world as it is matters in and of itself. It is the answer you might expect of a scientist. In fact it was given, in deed not word, by an artist, the Flemish Renaissance painter Jan van Eyck. There are five pictures attributed, with differing degrees of certainty, to van Eyck in which a realistic Moon adorns daylit or twilit skies. The clearest, and most moving, is a painting of the crucifixion dated between 1420 and 1425. A gibbous Moon sits low in the afternoon sky. The dark patches of the maria are clearly visible, as is the fuzziness of the night-edge.

One of the things that set van Eyck apart from his contemporaries was a commitment to recording the specific contents of his world as they were, even when such detail was incidental. The textures of his limestones show realistic weathering patterns; his mountains are topologically precise; his clouds pass meteorological muster. His Moon is, like them, a thing in the world, rendered not as allegory or icon but as it really is.

Not, though, as it really is in the afternoon. A gibbous Moon visible in the afternoon—the time of the Christ's death—would have to be waxing, not waning like van Eyck's.* This lapse seems to attest van Eyck's lack of interest in the Moon as an astronomical object; he just wanted to record how it looked. He probably made a preparatory sketch in the morning, when waning gibbous Moons set, and worked

---

* The full Moon always rises around sunset. Waxing gibbous moons, which can be seen in the days leading up to the full Moon, rise before sunset. Waning gibbous moons rise after sunset and as a result can be seen in the morning sky after sunrise.

from that. If what matters to you is capturing the look of the thing in itself, Moon is a moon is a moon.

Why a moon at all? Perhaps, as it is for many of us, it was simply a sight that he liked. Perhaps he saw it as a technical challenge. There may be another reason, though. The Moon had long been associated with death. In "Concerning the Face in the Disk of the Moon", Plutarch talks of the "substance of the soul being left on the Moon", where it "retains certain vestiges and dreams of life". He called the unseen farside the "Elysian Plain", the nearside the "Plain of Persephone Antichthon". Souls could travel from one side to the other through long "gulfs", or suffer in the "Chamber of Hecate" (probably *Mare Imbrium*).

Less learnedly, the Moon has the paleness of a skull, and dark sockets for eyes. In van Eyck's crucifixion, it is about the size of a skull, too, close to—and about as big as—the head of the impenitent thief to the Christ's left. Read like this, the painting inevitably brings to mind the greatest moonrise picture of the twentieth century, "Moonrise, Hernandez, New Mexico", by Ansel Adams, in which the brightness of the Moon in the evening sky matches that of the still-sunlit crosses in the high-country cemetery below.

Despite this echo across the centuries, van Eyck did not start a trend. The next artist to draw the Moon with features, as far as is known, was Leonardo, who sketched it with his unpublished notes on earthshine. But he saw no need to include its face in any finished art. The only other extant images of the Moon's features made before the advent of the telescope were made by William Gilbert, who was physician to Queen Elizabeth, and by Adam Elsheimer, a German artist living in Rome.

Gilbert's drawing, which dates from around 1600, is neither a sketch nor a work of art. It is a quite crude map, drawn on a grid, with various maria clearly delineated and rather prosaically named ("Southern continent", "Northern island", "Middlemoon Sea", etc). It is something which could, in principle, have been drawn at any time—had anyone wished to inspect the Moon that closely and record what they saw in a way that would allow others to refer to the same features. Before Gilbert, it seems no one did.

Gilbert's unprecedented interest in the Moon's face was like that of Galileo, Maestlin and Kepler in its ashen light: he, like they, was one of the small band that believed the Earth and the Moon were both bodies in motion. He had concluded that the Moon and the planets were not fixed in spheres, as Aristotelians like Albertus Magnus had taught, but that they shared a material, knowable and changeable nature with the Earth. He suspected, correctly, that the Moon rocked slightly in its orbit, sometimes showing the Earth a little more of its eastern hemisphere, sometimes of its western—a change which, in and of itself, would have been enough to give the lie to the idea of it being held fast in a crystal sphere centred on the Earth. It was to look for that change and others that he mapped the Moon; he was interested not in its features but in the chance that those features, or the perspective from which they were seen, might alter over time. Indeed, rather than expressing pride at being the first to map the Moon, he lamented that no one had done so before. If they had, past changes in its aspect might have been discoverable.

His way of seeing the universe provided Gilbert with a new reason to look closely at the features of the Moon. The same, I think, is true of Elsheimer, whose full Moon in "The Flight into Egypt" (1609) is the first since van Eyck to have distinctly dark maria and bright highlands. Elsheimer was associated with the thinkers of Rome's Accademia dei Lincei, who prided themselves on an intense interest in the details of the natural world (they took their name from the lynx because of its famously keen eyes) and whose number would later include Galileo. Some have suggested that Elsheimer's painting was based on observations made through Galileo's telescope—or perhaps even someone else's—but the case is weak, at best. However, if not telescopic, it is, like Gilbert's map and Galileo's observations, Copernican in intent: it clearly sees the Moon as a worldly thing, not a celestial orb.

It was a way of seeing that was soon to spread far beyond a small and discreet coterie of scholars. Looking at the Moon was about to be confirmed as a new way of seeing the world.

◗ ○ ◗

GALILEO WAS NOT THE FIRST PERSON TO SEE THE MOON THROUGH a telescope. But "The Starry Messenger" made him the person whose telescope changed the way that others saw it. As with the question of earthshine, it was his painterly attention to illumination that mattered: specifically, to shadow.

It is often said that Galileo's telescopic studies of the Moon revealed that it had features like the Earth's. John Milton, who met Galileo in 1638, said as much in "Paradise Lost", in which the "Tuscan artist", looking at the Moon "through optic glass", seeks

> *to descry new lands,*
> *Rivers or mountains in her spotty globe.*

But this is not quite right. Galileo did not find that the Moon had features like the Earth's. He found that the Moon, like the Earth, had features: it was the physical fact of those features' existence, not any particularly Earth-like character they might have, that mattered.

One indication of this is that Galileo showed no interest in mapping the Moon. He only showed its partial disk obliquely lit; he never named any features. His aim was to demonstrate the existence of distinct highs and lows on its surface, and his drawings and his analysis thus centre on the night-edge, where the rising or setting Sun provides the shadows that make relief most obvious.

Because the Moon's relief is dominated by craters, they dominate Galileo's analysis. He draws the reader's attention to the way that, at the night-edge, their rims, lit by the Sun, produce horns of light that pierce the night. His drawings exaggerate the effect substantially, at the same time making the craters larger than they in truth appear; their aim was to instruct the eye, not to mimic the sky.

In his text he laid particular stress on the way the darkness within some of the cavities persists into daytime, showing them to be low-lying. It is like early morning in the mountains, he says, when the Sun lights the western side of a valley first, its illumination subsequently sliding down to the valley floor and only reaching the eastern slopes when the Sun is quite high in the sky.* But Galileo did not say the craters were valleys or their rims, mountains: he preferred terms with less of the landscape about them, such as *prominences* and *cavities*.

Nor did he say that the "great and ancient spots" now called maria were seas. He said only that the shadows showed them to be smoother than the brighter surface, "like frosted glass", in which they were embedded, and also to be lower.† But that did not show they really were seas, just that, if anyone sought to "revive the old opinion of the Pythagoreans, that the Moon is another Earth . . . the brighter portion may very fitly represent the surface of the land, and the darker the expanse of water."

This part of his analysis set Galileo against others, such as Gilbert and Leonardo, who when seeing the Moon as a world in the sky had believed that it was the bright bits which corresponded to seas. Galileo saw them to be wrong. To see the sea as bright is to be dazzled, often literally so, by specular reflections. Look sunwards out to sea and the shifting surface provides a million mirrors aimed straight towards you; images of the Sun distinct, perhaps, close to merge into an ever more coherent band the further your eye rises towards the horizon. To either side of the path, though, the surface is darker, reflecting only empty sky. Seen from above and from a distance it is this darkness that dominates.

The passage in which Galileo makes this point seems to me one of the most extraordinary in the whole work: "I have never doubted that if the sphere of the Earth were seen from a distance, when flooded with the Sun's rays, that part of the surface which is land would present itself to view as brighter, and that which is water as darker in comparison." What sort of person, then, there, would have troubled himself to

---

* Though the darkness of the eastern slopes gets less intense the while, thanks to the secondary light from the bright-lit west.

† Until not that long before, it had been held that away from the shore, the Earth's great ocean was higher than the land—hence the phrase "high seas"—but by Galileo's day lowliness seemed strongly to suggest wateriness.

think about the Earth seen from a great distance—never mind reaching undoubted conclusions as part of the exercise?

Galileo taught his readers to see the Moon's third dimension as evidence that it was made of mundane matter—but not that it was, in fact, another Earth, with Earth-like features. On that he presented himself as agnostic. Having demonstrated the Moon's mundane nature, "The Starry Messenger" and its author moved on to other things. Galileo never published any further observations of the Moon.

The first part of the lesson was quickly heeded. Thomas Harriot and William Lower, two Englishmen, had both looked at the Moon through telescopes before they read "The Starry Messenger" but had not grasped what they were seeing. Describing his observations to Harriot, Lower wrote: "In the full she appears like a tart that my cooke made me last weeke; here a vaine of bright stuffe, and there of darke, and so confusedlie all over." After reading Galileo, they knew how to see what they were looking at: heights and cavities, roughness and smoothness.

As quickly as people learned to see the Moon as Galileo saw it, though, they also started to draw and talk about it in ways that he didn't—interpreting its features according to Earthly analogues, naming them and marking them down on maps. And here they ran into problems. Moonlight can be deceptive—even contradictory.

The fine shattered rock of the Moon's surface contains lots of small bits of glass that prefer to reflect light back in the direction it came from rather than off to one side; the same retroreflective effect is used in making cinema screens. This is why the Moon is much brighter when it is full, or close to full, than at other times. It is not just that more of the surface is lit; the over-your-shoulder way in which the surface is lit makes it more reflective than it is at other times.

These retroreflective particles are seen all over the Moon; but their distribution is not even, and its pattern does not reflect the underlying relief. The "rays" which extend out from young craters like Tycho, and which I take to be Lower's "vaine[s] of bright stuffe", are particularly densely populated with such particles and are thus striking when the Moon is full. But they are hardly discernible at other times. They are purely superficial features—not ridges or gullies—and thus cast no shadows. They have no more relevance to the shape of the surface they sit on than the trace of a kiss has to the contours of a cheek.

This is why the Moon looks so different when shadowless and full to the way it looks at its obliquely lit quarters. Mappers had to reconcile these aspects to produce images which both showed the relief imputed from shadows and did justice to what the full Moon looked like to the eye, and no two of them did so in quite the same way. The Moon that in the sky looks the same to everyone became idiosyncratically individualised when transferred to the page.

Then there is the matter of nomenclature. To begin with, different astronomers named the features on these maps according to different schemes. But the naming system used today was established in "Almagestum Novum" (1651; "The New Almagest") of Giovanni Riccioli, a Jesuit. Although the church had reservations about the physical truth of Copernicanism—and, more profoundly, about the idea that the authority to define truth in such matters should rest in other hands than its own—it also had many fine astronomers.

Riccioli, like almost every observer after Galileo, interpreted the darker parts of the Moon as watery, distinguishing them with the terms *mare* (sea), *oceanus* (ocean), *sinus* (bay), *lacus* (lake) and *palus* (marsh), and added to them terms associated with either seas, the Moon or both. Thus, as well as an ocean of storms and a sea of tranquillity, there is a bay of dew (*Sinus Roris*) and a bay of rainbows (*Sinus Iridum*), a marsh of decay (*Palus Putredinus*) and a lake of dreams (*Lacus Somniorum*).* Bright regions, meanwhile, were named for landforms—but only those defined by their relation to the sea: *terra*

---

* The International Astronomical Union, which takes care of such things, codified the remit of sea naming to characteristics of water and states of mind. When the Soviet Union was adamant that one of the seas its Luna 3 probe discovered on the far side was to be called *Mare Moscoviense*, diplomatic ingenuity allowed it on the basis that "Moscow is a state of mind".

(land), *littus* (shore), *insula* (island) and *peninsula*. This aspect of his nomenclature did not persist.

Most striking, though, were the names he gave to craters. In effect he turned his book's bibliography into his map's gazetteer: a list of astronomers and philosophers, ancient and modern. Kepler, whose discovery that planetary orbits are elliptical, not circular, laid the ground for Newton's theory of gravitation, got a fine bright-rayed crater. Tycho Brahe, the Danish astronomer whose meticulous observations allowed Kepler's discovery, got an even brighter one.

Tycho had come up with the astronomical system that Riccioli, and the church, favoured at that time—one in which the Moon, the Sun, Jupiter and Saturn revolve around a stationary and central Earth, while Mercury, Venus and Mars revolve around the Sun. This accommodated Galileo's discovery that Venus, like the Moon, had crescent and gibbous phases and that they waned and waxed in a way that could only be explained if the planet was circling the Sun. Copernicans took this as evidence that planets in general orbited the Sun. Tycho's system allowed the observation to fit into a world where, though some planets did so, the Earth remained central and stationary.

Despite favouring Tycho, Riccioli still gave Copernicus a splendid crater of his own. Indeed, the astronomer Ewen Whitaker, whose work on the history of lunar mapping is invaluable, suggests that the prominence given to Kepler, Copernicus and, most notably, Aristarchus—the ancient Greek who first suggested that the Earth circled the Sun—in Riccioli's scheme reflects a closet Copernicanism, one that he could not avow in the text but could at least hint at in his map.* This is, at best, a hunch. And even if taken at face value,

---

* What of Galileo? Riccioli gave him a prominent feature, too, but later observers found that the feature named for him was not, in fact, a crater, but a ray-like patch of peculiar brightness: it is now called Reiner Gamma, in accordance with later rules of nomenclature. By the time this had been realized, all the big craters were named; the crater now called Galilei is anomalously, even embarrassingly, small. In recompense, though, the four largest moons of Jupiter, which Galileo discovered and named after the Medici family to gain patronage, are now known as the Galilean moons.

with no hidden agenda, the map shows how revolutionary the times were. An astronomer had taken it upon himself to name the features of the Moon not after great nobles or churchmen but after scholars like him: a bold claim for the authority of knowledge, properly acquired. And that claim is particularly important in light of the fact that many of those scholars, and most of those who got the most spectacular namesakes, were moderns, not ancients. Riccioli's Moon was a celebration of new learning by new people.

● ○ ◐

WHILE RICCIOLI WAS PEOPLING MOON MAPS WITH ASTRONO-mers, others were filling the Moon itself with Moon people. A principle of divine economy—God would not be wasteful in his creation—led many to conclude that, if there were other Earths, there must be life on them. This is the burden of "The Discovery of a World in the Moon", published in 1638 by John Wilkins, an attempt to prove the Moon both habitable and inhabited. If they were inhabited, they must have stories. Thus for Copernicans and their fellow travellers, fictions of the Moon became, like drawings of the Moon, a way to expand on its worldliness, while also demonstrating the planetary nature of the Earth, waxing and waning in the lunar sky.

The wonder of earthlight, a rich signifier of the Earth being to the Moon what the Moon was to the Earth, was a frequent theme. Kepler talks about it in his "Somnium" (1634; "The Dream"), where earthlight mitigates, on the nearside, some of the climatic harshness resulting from 14-day days and nights. In Francis Godwin's "The Man in the Moone", published the same year as Wilkins's book, the protagonist, Gonsales, finds that most lunar life takes place lit only by earthshine; when Earth and Sun are both in the sky, the world is too much for all but its largest and noblest inhabitants; the rest all sleep through the long days.

The people on Godwin's Moon are large, wise and Godly, a great relief to him; when he cries out "Jesus Maria", the natives fall to their knees, and he rejoices at their kinship in Christ. An Earth-like Moon in a Godly universe, as Adam Roberts points out in his "History of Science Fiction" (2016), raised pressing new questions: are its inhabitants saved? Can they be? What does Christ's sacrifice mean in a universe far grander than that of the schoolmen? Must the people of the Moon take Communion?*

It was possible that the pure souls involved might be human—that the Moon might be materially the same as the Earth, as the Copernicans writing about it believed, but on a higher spiritual plane. Wilkins suggested that the Moon might be a "Celestiall Earth, answerable, as I conceive, to the paradise of the Schoolmen . . . this place was not overflowed by the flood, since there were no sinners who might draw the curse upon it". Cyrano de Bergerac, on visiting the Moon, found Eden to have been transported there, lock, stock and apple. He rather lowered the tone by making off-colour jokes about his trouser serpent to his local guide, the prophet Elijah.

There are echoes of Plutarch and his Moon full of purified souls here, and echoes of the association with death, too. There is also, though, the added Christian complication that a world of, or for, pure souls is to be found in the prehistoric past as well as hoped for in one's personal future. Thus, early in the history of stories about the Moon as new world, it becomes an old world, too. This dichotomy—ancient in itself, new to humanity—has remained a part of lunar fictions from then until now.

These contradictions of ancient and modern draw the Moon into a set of Earthly writings. It is not the only "other Eden, demi paradise" of the era's imaginings; European voyages of discovery regularly read strange islands as new Edens. To sail the oceans was to travel from the used-up and known to the new and unspoiled: to the "wooded island"

---

* If so, they may have been in trouble, as Johann Andreas Schmidt, a Lutheran theologian, argued in "Selenitas e luna proscriptos divini numinis gratia" (1679; "Selenites, or the Moon proscribed by Divine Grace"). The Moon, he noted, was too inclement for vineyards, and without vineyards there could be no Communion, and thus no salvation. There could thus be no people with souls on the Moon, merely monsters.

of Madeira or the "fortunate islands" of the Canaries, green and garden-like. It is largely as a variation on such an island theme that the Moon enters the literature of the age, part of a more general literature of fantastic voyages and of unusual isolates of humanity—a literature that includes Thomas More's "Utopia" (1516) and Shakespeare's "The Tempest" (c. 1610)—places of perfection and strangeness of magic and malformed mooncalf natives. In "Somnium" the Moon is explicitly referred to as an island; Godwin's Gonsales is an explorer who gets to the Moon from Saint Helena, at the time England's byword for the Edenic. His book's influence is seen in subsequent strange-islands-as-inquiries published by Defoe, Swift and others.

It is worth noting that in all these stories it is people from the Earth who go to the Moon. For most of the planet, the 17th century was not an Age of Exploration but rather an Age of Being Explored by Europeans. But it was the Europeans who were writing Moon stories, and few imagined themselves on the receiving end of exploration. When they did, centuries later, it would be to Mars, not the Moon, that earthlings would look for invasion.

The Moon thus became the farthest skerry of an archipelago of deep thought and high jinks, a place it held for centuries. These were not realistic fictions. Whether it was indeed habitable and what it might really be like, the questions which exercised Wilkins and Kepler, were, within a few decades, of relatively little interest. The Copernican revolution which Wilkins had been prosecuting in prose had by then been won by other means. By the end of the 17th century, the manner in which the Moon represented a new way of seeing the cosmos had changed from being a question of what it might be to live there, or how it looked, to a matter of the force that governed its movement.

● ○ ◑

Isaac Newton's "Principia" of 1687 tied the Moon to the Earth not by similarity but by gravity. Kepler had discovered that the planets moved in ellipses round the Sun; in the decades which followed it was confirmed that moons did the same around the planets fortunate enough to have them. He also found that orbiting objects moved faster when closer to the object they were orbiting than they did when farther away in a mathematically well-defined way. Working from these empirical laws, Newton produced a theory of how mass was attracted to mass—Earth to Sun, Moon to Earth, apple to Earth—in a way which depended on the square of the distance between them, a universal gravitation which, like the sight of the Moon from a kitchen window, unites the cosmic and the domestic.

This marriage of heavens and Earth was nowhere more visible than in the tides. To understand how gravity produces them, imagine an Earth completely covered by water. The water level directly under the Moon will be higher than average because the waters there, being closer to the Moon, are more strongly attracted to it than are the waters elsewhere: it is pulling them away from the Earth.

Somewhat counterintuitively, the water level on the side of the Earth directly opposite the Moon is similarly raised. This is because those waters, being farthest from the Moon, are the least strongly attracted to it—but the attraction they are feeling the least of is one that, in this geometry, pulls them *towards* the Earth. The strongest force pulling away from the Earth and the weakest force pulling towards it thus have much the same net effect. The result is something like a spherical volleyball Earth encased in a Moonwards-pointing rugby ball of water.

The Sun, too, creates tides, in much the same way: there is a bulge to sunward and another to anti-sunward. They are, though, smaller bulges. Though the Sun is 30 million times more massive than the Moon, it is also 400 times farther away, and the way that Newton's gravity works means that the distance counts against it more strongly than the mass counts in its favour.* The two bulges line up when the Earth, Sun and Moon do—that is, when the Moon is either very close

---

* Though gravitational force falls off with the square of distance, tidal effects fall off with the cube of distance, and 400^3 is more than 30 million.

to the Sun in the sky or directly opposite it. These are the spring tides associated with full Moons and new Moons; the way in which Newton's theory explained both their existence and their amplitude was one of the most impressive ways in which it brought the universal down to Earth.

The tides are in practice much more complicated than this. The Earth and its waters rotate every day; the point beneath the Moon makes a circuit only once a month; the point beneath the Sun only once a year. Thus the waters are endlessly trying to keep their tidally ordained shape while the solid Earth turns within it. This turning topography of shores and seabed means that the reach, frequency and precise timing of the tides vary from place to place. In the middle of the Pacific, with no local land to mess things up, the tidal reach is less than a metre. In the English Channel, where the bulge is trying to get from the Atlantic Ocean to the North Sea every 12 hours, the reach can be seven metres or so.

To Newton's irritation, his theory did not at first allow all these subtleties to be worked out from first principles; for a century or more, tide tables continued to be calculated empirically. But gravitation did mark a decisive shift to a worldview in which the universe had the characteristics of mechanism, in which Copernicanism was unavoidable and in which learning had new power. In so doing it also unburdened the Moon of the requirement to be Earth-like it had laboured under when it was part of the basic case for the Earth's planetary nature. The Moon was free to be what it was, and what it was came to seem ever less hospitable.

The changes to the Moon's face that Gilbert had planned to record turned out to be, as he had suspected, slight changes in which parts of its surface were visible, changes that Kepler's laws and Newton's theory explained precisely. But there were no changes to be seen in what was on that surface. There were no weather patterns that altered with the seasons. There was no snow. The "seas", on close inspection, turned out to be peppered with small marks, and not completely flat. Odd water, to say the least.

Nor was there air. The idea that the Moon had an atmosphere its inhabitants could breathe had been crucial to Wilkins's arguments for a "world in the moon". Indeed, his book is the first to ever use the word *atmosphere* in its modern sense. There had been no atmosphere on the Earth before: just air, an element that sat above the land and sea. Only when it became a necessary condition for life on another planet did air develop a planetary mode of being as an envelope that could be wrapped around any body of sufficient size. Only after atmospheres became a way of understanding air elsewhere did they become a way of understanding it on Earth, too.

Unfortunately, Wilkins's evidence for the Moon's atmosphere—the blurriness of features on the lunar surface—was in fact evidence of the Earth's. By the late 17th century, studies of what happened when the Moon, moving across the sky, obscured a distant star seemed to prove this. The star would not fade or flicker as it approached the limb of the Moon, as it might have done were it being seen through ever more of the Moon's obscuring atmosphere. It simply vanished.[*]

The lack of an atmosphere confirmed what observation was already making clear about the maria: they were dry. One of the key discoveries of knowledge's mechanistic turn was that though nature might, as Aristotle was held to have taught, abhor a vacuum, machines—"air pumps"—could create one. When they did so, liquids under the vacuum boiled away, whatever the temperature. An airless Moon must be a sea-less one, too. It thus seemed an ever more lifeless one, to boot.

None of this devalued the Moon as a site for speculation and Swiftian satire. It did, though, provide a new object for that satire—the fanciful scientist attending to the high-flung matters of the Moon and

---

[*] Later astronomers took this argument further, arguing that with a thin enough and stable enough lunar atmosphere, the star being transited would not be distorted, but its apparent position might change because of its light being refracted through that atmosphere. Sir George Airy, among others, looked for this phenomenon—and thus developed techniques which would in time be used to look for the effects of general relativity during solar eclipses and held up as the first confirming evidence of Einstein's theory of general relativity.

ignoring the everyday. Thus in Samuel Butlers's "The Elephant in the Moon" (c. 1670), astronomers mapping the lunar surface

> *To make an inventory of all*
> *Her real estate, and personal;*
> *And make an accurate survey*
> *Of all her lands, and how they lay,*
> *As true as that of Ireland, where*
> *The sly surveyors stole a shire*

are amazed to discover on its face not merely armies of tiny beings but also a very big elephant:

> *It is a large one, and appears more great*
> *Than ever was produc'd in Afric yet;*
> *From which we confidently may infer,*
> *The Moon appears to be the fruitfuller.*

As they rush off to right their papers, though, a servant learns the truth when he looks not through but into the telescope.

> *He found a small field-mouse was gotten in*
> *The hollow telescope, and shut between*
> *The two glass-windows, closely in restraint,*
> *Was magnified into an Elephant.*

The armies were gnats, the mighty monster lowly vermin; instead of astronomers turning the cosmos upside down with their instruments, the social order is overturned by smart servants with foolish masters. A similar sentiment shapes Aphra Behn's play "The Emperor of the Moon" (1687), in which Doctor Baliardo refuses to let his daughter and niece marry men from Earth because his vast telescope has revealed to him, he thinks, that the men of the Moon are more advanced, and thus preferable. He has clearly read Lucian, Wilkins, Godwin and more besides, but learning has not made him wise: "Lunatick we may call him without breaking the Decorum of good Manners," says the servant Scaramouche, "for he is always travelling to the Moon."

The young women enrol Scaramouche and his fellow skivvy, Harlequin, in a subterfuge to convince Baliardo that their suitors, Don Cinthio and Don Charmante, are just the sort of Selenites, as Moonpeople are often referred to, that he wants them wooed by—indeed, that they are the Emperor of the Moon and his brother, the King of Thunderland. Kepler and Galileus are brought on as character witnesses, and much hilarity ensues—as does great spectacle, with stage machinery as advanced as any of the time.

The possibility of taking the audience to the Moon has since often led to new excesses and subtleties of the spectacular. Georges Méliès's "Le Voyage dans la Lune" (1902; "A Trip to the Moon") used the new magic of cinema to show the Moon as a realm unlike any other; "A Trip to the Moon", a literally sensational fairground ride that opened in 1903 on Coney Island, led to fun fairs and amusement arcades across the world becoming known as Luna Parks; Kubrick's "2001: A Space Odyssey" (1968), with its anodyne airline elegance settling weightless into savage emptiness, set the style for later special effects extravaganzas while outdoing them in substance.

That "The Emperor of the Moon" led the way in uniting the mechanisms and spectacles of the stage and the mechanistic spectacle of nature was not a coincidence. Behn had recently translated into English a very popular French book on worlds beyond, Bernard le Bovier de Fontenelle's "Entretiens sur la Pluralite des Mondes" (1686; "Conversations on the Plurality of Worlds"). In a series of dialogues between a philosopher and a noblewoman, held in a moonlit garden, Fontenelle lays out his ideas of a Copernican cosmos and its possible inhabitants. The idea that wonders of the world have hidden mechanisms, whether in the observatory or on the stage, and that few have the expertise necessary to see them—Doctor Baliardo certainly doesn't—is a theme in both works. As Fontenelle's philosopher puts it to his marquise:

Nature is a great Scene, or Representation, much like one of our Operas; for, from the place where you sit to behold the Opera, you do not see the Stage, as it really is . . . the wheels and weights which move and counterpoise the Machines are all concealed from our view. [T]here is not above one Enginier in the whole Pit, that troubles himself with the consideration of how those nights are managed. . . . You cannot but guess, Madam, that this Enginier is not unlike a Philosopher.

● ○ ◐

"MICROGRAPHIA" (1665), THE FIRST BOOK PUBLISHED BY THE Royal Society, is famous for its exploration of hidden worlds through its spectacularly rendered images of life as never magnified before: a grain of pollen, a fly's wing, a louse reproduced at the size of a page. But its author Robert Hooke—an engineer not unlike a philosopher, enemy to Newton and friend to John Wilkins—was interested in magnifications of worlds larger and more distant, too. "Micrographia" contains an image of the crater Hipparchos and its surroundings, remarkable both for its extremely fine detail and for being one of the first images of a specific feature on the Moon rather than of the Moon as a whole.

The specificity comes from Hooke's interest in craters not as evidence of the Moon's imperfect nature but as phenomena in themselves: what, he wondered, created such things? Kepler had suggested that they might be great circular barrows: dwelling places built up by Moonpeople to let them move from the shadows in the east in the long mornings to the shadows in the west in the long afternoons. Hooke saw them instead as expressions of movement not across the surface but through it: of either something going in or coming out.

Through experiment, he found that if he "let fall any heavy body, as a Bullet" into "a very soft and well temper'd mixture of Tobacco-pipe clay and Water . . . it would throw up the mixture round the place, which for a while would make a representation, not unlike these of the Moon." Alternatively, if "boyling Alabaster . . . being by the eruption of vapours reduc'd to a kind of fluid consistence . . . be gently remov'd besides the fire . . . the whole surface, especially that where some of the last Bubbles have risen, will appear all over covered with small pits, exactly shap'd like these of the Moon."

It was for the second, internal, cause that he plumped. The splash-craters made by dropping things into watered-down pipe clay were transient; the bubbled-up pockmarks on the alabaster remained after it set solid: a mark in their favour. The details of their appearance in oblique light was another such mark. "By holding a lighted Candle in a large dark Room, in divers positions to this surface, you may exactly represent all the Phænomena of these pits in the Moon, according as they are more or less inlightned by the Sun."

Hooke was aware that his analogues might mislead. But what his models suggested, analogues confirmed. Volcanoes such as those of Iceland, the Canaries and New Spain had tops shaped "like a dish, or bason" and raised "great quantities of Earth" around them, which he felt bolstered the case for an internal origin. As to the case for bullets, he saw astronomy as revealing nothing which might hit the Moon.

Hooke's volcanic account of the Moon held the field for most of the next three centuries, reaching its most fully theorized, and most beautifully illustrated, form in the works of a Victorian industrialist, James Nasmyth. Nasmyth was, from an early age, much taken with volcanoes. On walks around their home city of Edinburgh, his father, a noted landscape artist, had taught him that they were responsible for its dramatically knobbily topography. His father's friend and benefactor, Sir James Hall, after whom Nasmyth *fils* was named, was one of the first men of science to try to emulate volcanoes by heating rocks to their melting point to see what lavas he could make.

Most of Nasmyth's meltings, at least in his professional life, were in foundries; but the link to the world of rocks was not remotely lost on him. In 1840 he took a break from a tour of various European shipyards and arsenals which employed steam hammers and other machines of his company's manufacture to climb Vesuvius. Having marvelled at the vent at the centre of its crater from as close a vantage point as he dared, he later recalled, "I tied the card of the Bridgewater Foundry to a bit of lava and threw it in, as token of respectful civility to Vulcan, the head of our craft." When he retired rich at the age of 48, he devoted himself to the studies of the Moon that he had previously undertaken in his spare time. He looked forward to "the tranquil enjoyment which results from the study of one of the Creator's most potent agencies in dealing with the materials of His worlds, namely, volcanic force."

"The Moon, Considered as a Planet, a World and a Satellite" (1874), the book that Nasmyth, aided by his friend the astronomer James Carpenter, produced from these studies, worked methodically through the three characterisations of its subtitle. It was the first—the Moon's nature as a planet—that fascinated Nasmyth most.

As the astronomer Richard Proctor wrote at roughly the same time, the issues "of progress, development and decay" were "the principal charm of . . . all observational science". Proctor saw the lack of such changes as a deficit on the Moon's part. Nasmyth and Carpenter thought them discoverable with the concepts that Victorian science had developed to understand progress, development and decay: evolution—a concept which included, but went beyond, Darwinian natural selection—and thermodynamics, the nascent science of energy, heat and work. The now changeless Moon had previously undergone "a constant progression from one stage of development to the next . . . a perpetual mutation of form and nature"; it had evolved. And an understanding of that evolution could explain the history of other planets, most importantly the Earth, in new ways.

The book follows the account of the solar system's origin given by the great French astronomer Pierre-Simon Laplace: a nebula of dust and gas collapsed in on itself because of Newtonian gravitation. As it did so, an enormous amount of potential energy was given up, and the first law of thermodynamics stated that that energy could not just vanish. Instead, it went into heat. As one of the first law's framers, Julius von Mayer, had put it, the nebula's collapse was a source of heat "powerful enough to melt worlds". Nasmyth and Carpenter believed the Earth and Moon to have been born molten and to have solidified from the outside in.[*]

The Moon cooled from its molten state quicker than the Earth. This was because it was small, and small things cool quicker than big ones, and because it lacked an atmosphere, and so was not kept warm by the greenhouse effect.[†] Nasmyth's foundry experiences convinced him that the solid crust must have been less dense than the liquid beneath it, and so as the crust grew it squeezed the molten layers beneath it. Eventually, the pressure below became so great that the crust could no longer contain it; molten lava rushed up to the surface and out into the void beyond it. Extreme pressures, low gravity and a lack of any air resistance provided "conditions most favourable to the display of volcanic action in the highest degree of violence."

The lava in these eruptions did not just flow down the sides of mountains. It flew tens or hundreds of kilometres into space before falling back to the surface. The eruptions were like parabolic fountains in which water is squirted up from the middle of a pool and comes back down all around its rim. They thus created great circular walls, not single peaks.

This theory explained why, whereas the craters of Earthly volcanoes sat atop mountains, the interior of lunar craters often sat lower than the surrounding plains: as vast amounts of hot rock from the depths shot out into space, the undermined surface subsided as the lava built up all around. It also explained why many craters had solitary peaks at their centre: they were the last gasps of the eruption, no longer able to throw lava tens of kilometres, but still able to build up a mountain in the more subdued Earthly style.

Understanding the Moon's surface as a function of its history in this way provided a new way of thinking about the Earth. Geology was at the time a "uniformitarian" discipline: it insisted that the past

---

[*] The 21st-century account of this origin, complete with melted worlds, is found in Chapter IV.

[†] Already an accepted notion among scientists of the time, though not under the same name as today.

had been sufficiently like the present that through an understanding of the processes in the world today—erosion, sedimentation, volcanism—you could explain all that needed explaining about the past. The cosmic perspective offered by the Moon suggested that the past could have been very different even though the Earth, the Moon and the solar system were all embedded in a cosmos changed by the working out of physical laws.

As well as a prototypical planet in the cosmos, though, Nasmyth and Carpenter also saw the Moon as a satellite of the Earth—a relationship which they believed to be defined by its utility. Considering the Moon as a satellite meant asking what it does for earthlings.

As a source of light, it gets short shrift. For most people, through most of history, this has surely been the Moon's great importance— bringing light to the night, and to some nights more than others. But to an up-to-date Victorian, his streets lit by gas and his house lit by paraffin, moonlight was not what it had been as recently as a century before, when an earlier generation of industrialists and inventors, the members of the Lunar Society of Birmingham, met at the full Moon because it made it easier to ride home. For Nasmyth, moonlight was all very well for poets, painters and peasants. Indeed, it should excite "our warmest admiration". But it wasn't really up to snuff: it was changeable, fugitive, partial, imperfect and of secondary importance. For men of action, tides were the measure of the Moon.

"Rest and stagnation are fraught with mischief," says the bustling Victorian businessman. "Motion and activity in the elements of the terraqueous globe appear to be among the prime conditions in creation." The Sun provides this highly desirable motion and activity by driving the winds. The Moon does the same for the waters, cleaning out the corruption of estuaries such as those of the Thames and Mersey. It is "our mighty and ever active 'sanitary commissioner'".

Tides provide not just cleanliness: they aid in commerce, too. The extra lift outgoing tides provide to ships and barges leaving port saves a city like London thousands of pounds a year, maybe millions. And in time they will provide ever more power. Britain's coal—"bottled sunshine", as the book puts it—is inevitably going to run out. The mechanical power of the tides, transformed into electricity and passed down wires to the industries that need it, could become the nation's new prime mover.

Light and sanitation are not the only satellite services; navigation and timekeeping get a look in, too. And despite Nasmyth's no-nonsense business sense, not all the uses of the Moon are utilitarian. There is something more elevated—its deathliness teaches us to esteem the habitable Earth even more highly. And it reveals the truth of an earlier age in a way nothing else can do. The foundryman's eye sees in it a planet "with all the igneous foundations fresh from the cosmical fire, and with its rough-cast surface in its original state, its fire and mould marks exposed to our view." The scientist's mind rejoices at the sight.

One of the two founding documents of British geology—and its uniformitarian mindset—was James Hutton's "Theory of the Earth" (1788). Those perambulatory discussions of the crags of Edinburgh would have made copious reference to the great man; Nasmyth's namesake James Hall had been Hutton's leading disciple. The most famous line from Hutton's "Theory", with its vistas of the untiring grinding-down of mountains and the slow resurgence of seabeds into the sky, is its last one: "The result, therefore, of our present enquiry is, that we find no vestige of a beginning, no prospect of an end." On the Moon, though, write Nasmyth and Carpenter, "its pristine clearness unsullied, every vestige [is] sharp and bright as when it left the Almighty Maker's hands." The Moon, frozen at an earlier stage of planetary evolution, reveals to its student what the eroded Earth cannot. The vestige of a beginning, lost on Earth, is preserved in the sky, ancient and new.

● ○ ●

THE BOOK'S IDEAS ARE FASCINATING; ITS LANGUAGE AND ASIDES are often a delight. But what Nasmyth's work is best remembered for is not its words or arguments but its illustrations: in particular, its photographs.

Almost none of them, surprisingly, are photographs of the Moon. Though astronomers were using photography by the 1860s, their pictures of the Moon were not terribly good, especially not for bringing out the features that Nasmyth cared about most. Yet photographs were setting a new standard for fidelity in representation, for recording the specific contents of the world as they were. They were of the moment. So Nasmyth photographed the Moon as he imagined it.

Nasmyth's father, having made sketches of landscapes outdoors, would retreat to his workshop and make clay models of what he had seen based on the sketches and his memories. His son thus learned to treat modelling as a way of thinking, one which placed the skill and experience of hand and eye at the service of reason and analogy. And that was how he made his Moon visible to the world. He sketched, he modelled, and then, having mastered the techniques of photography expressly for this end, he photographed the models in strong, oblique light rather like the one with which Hooke observed his cratered alabaster. In doing so, he probably influenced how the Moon was seen by others more than anyone else between Galileo and the 1960s.

The lunar drawings Nasmyth had made earlier in his career were both beautiful and widely appreciated—he showed them to Queen Victoria herself—but his modelled photographs had a sharpness that went beyond them, especially in the depth of their shadows. Many of the pictures look straight down onto specific features of the Moon. Some, such as his picture of Theophilus, Cyrillus and Catharina, a trio of similarly sized large craters which is thrown nicely into relief when the night-edge passes them about five days after the new Moon, can, at first glance, be mistaken for photos of the real thing.[*]

There are also pictures "taken" obliquely, or from the standpoint of an observer standing on a lunar plain. One shows the Sun eclipsed by

---

[*] Catharina is notable for being one of fewer than 30 lunar craters, out of over 1,600 with names, to be named in honour of a woman, in this case Saint Catherine of Alexandria.

the Earth, its atmosphere a fierce red ring. Another, switching points of view, shows the Bay of Naples in the same straight-down view as used in the pictures of the Moon, the better to draw out Nasmyth's analogy between its craters and those of the Campi Flegrei and Vesuvius (features which, it must be admitted, look a lot more lunar in Nasmyth's model than they really do from orbit).

A few move into pure analogy. Nasmyth's ideas about the wrinkling that the Earth's slower-than-lunar cooling imposed on its surface are strikingly illustrated with pictures of long-since-picked apple and an old man's hand (his own). A glass bulb cracked by a small expansion demonstrates the mechanism by which he believed the rays emanating from Tycho and other bright craters were made.

In Jules Verne's "Autour de la Lune" (1870; "Around the Moon"), the sequel to "De la Terre à la Lune" (1865; "From the Earth to the Moon"), Barbicane, the leader of the expedition, looks down from orbit in wonder at Tycho's rays. Michel Arden, a poet along for the ride, says that they look like cracks made in glass by a thrown rock—perhaps, in this case, by a comet. Barbicane pooh-poohs the idea; the force must have come from a shock within. This is not just obvious to his educated eye. It "is the opinion of the English savant, Nasmyth".

"No fool, that Nasmyth," replies the assenting poet.

This is just one of the ways in which the Moon Verne's travellers circle is Nasmyth's Moon; one of mountainous annular volcanoes, lofty impassable ramparts and little else. In particular: no air, no streams, no woods, no life. Verne's book is the first story of the Moon to find it uninhabited. It may have been inhabited once—Arden sees what he takes to be a ruined city and aqueduct—but now, the travellers agree, it is almost certainly dead.

What about the Moon is interesting, if the Moon is dead? Getting there. Verne's original book is the first Moon story in which the technology that gets explorers on their way is more interesting than their destination. The ability to reach the Moon, Verne says, marks the point at which humans have become a power of truly planetary significance. In a way Nasmyth would doubtless have appreciated (alas, I know of no record that he read the book he influenced), the casting of the great 900-foot cannon that sends Arden, Barbicane and Captain Nicholl to the Moon is explicitly compared to a volcanic eruption:

A savage, wandering somewhere beyond the limits of the horizon, might have believed that some new crater was forming in the bosom of Florida, although there was neither any eruption, nor typhoon, nor storm, nor struggle of the elements, nor any of those terrible phenomena which nature is capable of producing. No, it was man alone who had produced these reddish vapors, these gigantic flames worthy of a volcano itself, these tremendous vibrations resembling the shock of an earthquake, these reverberations rivaling those of hurricanes and storms; and it was his hand which precipitated into an abyss, dug by himself, a whole Niagara of molten metal!

The launch is grander still:

An appalling unearthly report followed instantly, such as can be compared to nothing whatever known, not even to the roar of thunder, or the blast of volcanic explosions! No words can convey the slightest idea of the terrific sound! An immense spout of fire shot up from the bowels of the earth as from a crater. The earth heaved up, and with great difficulty some few spectators obtained a momentary glimpse of the projectile victoriously cleaving the air in the midst of the fiery vapors!

The travellers, Verne tells us, have "placed themselves beyond the pale of humanity, by crossing the limits imposed by the Creator on his Earthly creatures." So, it seems, had the industry that made that transgression possible, a human power no longer distinguishable from the powers of the Earth itself.

●　○　◑

NASMYTH'S SENSE OF THE MOON AS A KEY TO UNDERSTANDING the Earth's evolution still rings true. So does his vision of planets which, like steam engines, are shaped by great flows of energy and the laws of work and heat those flows obey. He got some remarkable details right, too, suggesting that bacteria—then, in the form of "germs", a very new-fangled concept—might be able to survive in space when all other forms of life could not or that the appearance of the Earth from the Moon would be dominated by ever-shifting bands of cloud, not continents, or that tidal barrages could produce large amounts of electricity. But he got the Moon itself profoundly wrong, both in the general way that he explained it and in one specific way he envisioned it.

To take the second first: in order to produce the sort of shadows he saw through his telescope—and, one cannot help but think, to satisfy his own liking for the craggy and picturesque—Nasmyth made his lunar mountains very steep and jagged. In so doing he set the visual template for future renditions. The great 20th-century American astronomical artist, Chesley Bonestell, made his Moons just as craggy when he painted them for magazines like *Life*, *Scientific American* and *Collier's*, or as backdrops for the film "Destination Moon". So did those whose drawings drew on Bonestell's, as the Belgian cartoonist Hergé's did when he sent his boy reporter Tintin to the Moon. And why should the Moon's mountains not be craggy and magnificent? There were no winds, no rains, no glaciers to erode them. They should be raw and sharp.

But they are not. Nasmyth, and most of those who came after him, underappreciated the exaggeration that oblique lighting throws on quite modest relief.* The sharp shadows that people had been interpreting since Galileo are cast by soft, rounded things, far less angular than the Andes or the Alps. They are shoulders, not shards. There is barely a slope on the Moon that couldn't be walked or clambered up,

---

* The French astronomer, artist and illustrator Lucien Rudaux (1874–1947) is a notable exception.

even if the gravity was as strong as the Earth's. If the Moon sports anything that could be fairly called a cliff, it has yet to be seen.

This is because, despite the lack of wind, rain and ice, the mountains of the Moon are indeed being eroded. They are constantly hit by particles of dust moving at orbital speeds. The bombardment is not intense—1,800 tonnes of dust a year, even moving at over 10km/s, imparts far less energy to the Moon's surface than a few minutes' worth of rain does to the Earth's. But this thinner-than-thistledown, faster-than-bullets onslaught is unremitting. No crag can stand against it over a billion years.

This misapprehension as to erosion speaks to Nasmyth's greater, deeper mistake. The bombardment of the Moon is not limited to dust. It has included objects up to the size of pretty big asteroids. Such objects can deliver extraordinary amounts of energy, enough to reshape landscapes on the largest of scales in just a couple of seconds.

In 1941, while waiting for a lecture at the Field Museum in Chicago to get under way, a Midwestern industrialist who had been an astrophysicist in his youth, Ralph Baldwin, was struck by curious grooves visible in a photograph of the Moon. These were not superficial rays like those that emanate from Tycho but shadow-casting gouges that showed real relief. It looked, he thought, as though something had moved not into, or out of, the Moon but across it, at great speed and with great force. What process could create such a sculpture?

Intrigued, Baldwin studied other pictures and observed the Moon itself. He found that the gouges that had caught his eye had a radial structure, fanning out from a point in the centre of *Mare Imbrium*, and concluded that this "Imbrium Sculpture" had been created by debris thrown out by a giant impact. It followed from this that the arcing mountains that defined *Imbrium* were the rim of a crater—one which, at over a thousand kilometres across, would have dwarfed Tycho or Copernicus. The dark lavas that made *Imbrium* a sea were the result of eruptions that had come after the impact, filling up part of the great void that it left.

And if the great circle of *Imbrium* had been formed that way, then surely all the other lesser circles had, too. The craters of the Moon, from small to enormous, had all been formed by impacts.

This idea, which Hooke had decided against in part because he could not imagine what such impactors might be, had been around for a time. Proctor toyed with it in the 19th century. Grove Karl Gilbert, the 19th-century American geologist a little crazy on the subject of the Moon, believed that the Moon's craters and maria had been created when leftover building blocks of planetary formation, "planetesimals", had fallen on to it in its early years. In the early 20th century, Ernst Öpik and Charles Gifford, astronomers from Estonia and New Zealand, respectively, independently came to the same conclusion by considering the ever-growing number of asteroids and comets with which careful observation was populating the solar system.

But the idea did not really develop any traction. This may in part have been the result of bad luck and bias. Though Gilbert was famous, his Moon paper was published in a peculiarly obscure journal; astronomers did not look to Estonia and New Zealand for the next big thing, less still to the Oliver Machinery Company of Grand Rapids, Michigan, where Ralph Baldwin was to be found. But there was also a deeper reason for resistance. If the Moon was thus battered, its neighbour the Earth must have a catastrophic past, too.

Geologists, committed to their uniformitarianism, deeply disliked catastrophes. Astronomers didn't much care for them, either. When, in "Around the Moon", Barbicane gently mocks Arden for suggesting one of the "much abused comets" as a cause for the bright rays from Tycho, he was echoing Francois Arago, an astronomer friend of Verne and Nasmyth. Arago made it part of his mission in life to convince anyone who would listen that comets were not ill omens and should not be imagined as agents of collisional doom. They should be appreciated for the unthreatening wonders of the sky they were.

The person who changed this, thereby both creating the framework for the modern geological understanding of the Moon and playing a key role in the realisation that, in fact, impact catastrophes do take place

on Earth, was a geologist called Gene Shoemaker. Shoemaker had a
magnificent scientific imagination, a fierce commitment to expanding
geology's disciplinary reach beyond the Earth and a powerful person-
ality (his colleagues lampooned him in office-party skits as Dream
Moonshaker). He also had something else which proved crucial: experience of nuclear explosions.

Studies Shoemaker made of Jangle U and Teapot ESS, two craters
in Nevada created by underground nuclear tests, showed them to have
distinctive features unlike those of craters formed by volcanism, most
obviously curled rims in which the rock layers were folded back on
themselves. Working from an unpublished paper by Edward Teller,
Shoemaker began to understand the power and behaviour of the shock
waves that created these features.

It was with eyes thus trained that he visited, in 1957, another crater: Barringer Crater, also known as Meteor Crater, about 70km east of
Flagstaff, Arizona, and a bit more than a kilometre across. Its owners,
the Barringers, had believed for generations that it had been formed
by a large meteor—in part because many metal meteorites had been
found nearby—and had put considerable money and effort into trying to reach the body of meteoric iron they believed must be buried
beneath its floor. The Geological Survey had taken the position that it
was a "maar", a peculiar sort of volcanic crater created when subterranean magma vaporises an aquifer. The Barringers blamed the Survey's
mundane explanation for scaring off investment. As an employee of
the hated Survey, it took Shoemaker some time to gain their trust.

There was a striking irony in all this. The Survey scientist who had
earned the enmity of the Barringers by declaring the crater a maar
had been none other than Gilbert himself. Hearing of the crater and
its associated meteorites, and believing as he did that lunar craters
were created by impacts, Gilbert had been keen to investigate Meteor
Crater as a possible Earthly exemplar of that which he could normally
only study through a telescope. In Arizona, though, he found no evidence of an impact. There was no evidence of anything buried beneath the floor of the crater. What was more, the sediments that had

been thrown out of the crater onto the surrounding plateau seemed, he thought, to have the same volume as the crater itself. Where, then, were the remnants of the impactor? Much as he might have wanted to find an impact crater, he could not convince himself that this actually was one; it was a maar, and the nearby meteorites were a coincidence.

Shoemaker showed that Gilbert had been wrong about the impact for the same reason that the Barringers had been wrong to think there was a motherlode of valuable metal to be found beneath the crater. Meteor Crater had indeed been made by a piece of extraterrestrial metal, but there was more or less none of it left. A metal body just 50m or so across—very small compared to the size of the crater—had hit the plateau at a speed high enough to release about 10,000 times more energy than the small nuclear blasts at Teapot ESS and Jangle U. The energy was in the form of two shock waves: one slammed forward into the plateau's limestone; the other, backwards into the impactor. The forward shock scooped out Meteor Crater in just the same way that the shock waves from the nuclear detonations at the Nevada test site scooped out theirs, throwing a crater's-worth of ejecta out onto the surrounding plateau and leaving the lip of the crater turned over in the telltale way Shoemaker had learned to recognise. The reverse shock vaporised the impactor. Daniel Barringer had never found the crater-forming meteorite because it had blown itself up.

Shoemaker was not the first to see the analogy between the destructive power of the bomb and the cratered face of the Moon. Robert Heinlein was, along with Arthur C. Clarke, one of the two great Moon writers of the decades leading up to Apollo. The first of his lunar novels, "Rocket Ship Galileo" (1947) was also the first of the series he wrote to prepare post-war children and teenagers for the post-Hiroshima world. I bought it decades later, at the age of 10 or 11, at a church bookstall. It is the story of a nuclear physicist who recruits three all-American teenage boys to make the first trip to the Moon.

Ross floated face down and stared out at the desolation. They were . . . approaching the sunrise line of light and darkness. The

shadows were long on the barren wastes below them, the mountain peaks and the great gaping craters more horrendous on that
account. . . . "I'm not dead certain I'm glad I came."

Morrie grasped his arm, to steady himself apparently, but quite
as much for the comfort of solid human companionship. "You
know what I think, Ross," he began, as he stared out at the endless
miles of craters. "I think I know how it got that way. Those aren't
volcanic craters, that's certain—and it wasn't done by meteors. They
did it themselves!"

"Huh? Who?"

"The moon people. They did it. They wrecked themselves. They
ruined themselves. They had one atomic war too many."

"Huh? What the—" Ross stared, then looked back at the surface as if to read the grim mystery there.

Nuclear weapons gave Heinlein a way to imagine instantaneous energies creating craters many kilometres across, reshaping landscapes in
a fuckflash—the same insight Shoemaker's study would later confirm. He
was wrong to discount meteors. But he was right about the sheer and
sudden scale of the destruction. The energy that creates the craters of
the Moon is not the energy of the foundry. It is the energy of the bomb.

●   ○   ●

WHEN HE WAS A BOY IN THE 1950S, IN LOVE WITH BOTH SCIENCE
and art, Bill Hartmann, like James Nasmyth, used to make plaster
models of the Moon's craters to try to grasp what it would be like to
see them not from above but from the side. In the 1960s, as a graduate
student at the University of Arizona in Tucson, he worked on a way of
seeing that did the reverse, allowing bits of the Moon normally seen
obliquely to be looked on as if from above. It was a way of looking
that revealed the Moon to be not just subject to impacts but subject to
almost nothing but impacts.

Gerard Kuiper, Bill Hartmann's boss and a pioneer of planetary science, had procured a white hemisphere about a metre in diameter on to which he and his assistants could project telescopic images of the Moon's nearside. In general, if you project a picture onto a sphere, you will distort it; it will flow down the flanks like stretched toffee. But if the picture you have is of another sphere, and you get the projection just right, what was distortion becomes rectification. A two-dimensional image of one half of a sphere projected onto a hemispherical screen produces a three-dimensional image.

Kuiper's technique thus allowed him and his students to see the near side of the Moon from new angles: in left and right profile, as it were. Bill's job was to take pictures of the projection for a "Rectified Atlas of the Moon" that Kuiper was working on. Examining the Rook Mountains and, to the south of them, *Mare Orientale*, an obscure, oblique blob on the Moon's western limb as seen from the Earth, he had what he later called a "Eureka Moment".* The Rook Mountains were not just a line of peaks. They were one arc of a pair of concentric features that encircled *Mare Orientale*'s dark basalt heart; and the surrounding landscape was cut by a "complex mass of radial valleys and striations" coming from the centre of the sea like the sculpture which Baldwin, and before him Gilbert, had seen around *Mare Imbrium*.

That there were big impacts on the Moon had, in the few years since Shoemaker's breakthrough, become widely accepted. They explained arcuate mountain ranges like the Alps and Apennines around *Mare Imbrium*. Baldwin's idea that the dark maria were sheets of basalt

---

* Why is *Mare Orientale* in the west? On Earth, the direction in which the planet is spinning and the direction in which the night-edge is moving are opposed to one another; the planet spins to the east, and the night-edge moves to the west. On the Moon, the two directions are the same. Early astronomers defined the direction in which the Moon was spinning as being its east and, according to that convention, *Mare Orientale* is appropriately named. But that convention has the night-edge moving east, too, which freaks out today's Moon scientists and astronauts: the Sun really should not rise in the west. So, they flipped the definitions of east and west, leaving *Mare Orientale* ill named.

that had erupted into the basins formed by large impacts a lot later on—hundreds of millions of years later, it transpired—had become pretty much conventional wisdom.

What Bill discovered was that this explained more than the maria. There were ringed basins that had almost no basalt within them, too. The more Hartmann looked at the rectified projection of the Moon, the more such markings he saw. All the big mare basins had vestiges of such structures around them; and so did large basins that contained no basalt.

Some of what he was seeing had been described piecemeal before; but once Bill had learned how to look for basins, he saw the surface as a whole beginning to fit together in a new way—a new gestalt, as he would later put it. Impacts explained everything, from small bowl-like pits to simple craters a few kilometres across to larger craters with their distinctive central peaks to basins with multiple rings and more complex internal peaks, some of the biggest of which were filled with maria lavas. The maria were not uninteresting, but they were epiphenomena. All the large-scale structure of the Moon was down to impacts.

In subsequent decades "multi-ring impact basins" have proved to be a near universal feature of planetary surfaces. They have been discovered on all the rocky inner planets and most of the larger moons of the outer ones: Valhalla, on Callisto, is a particularly fine example. Venus, with a youngish crust, offers only a few. Mercury is covered with the things. The Earth's have mostly been erased by erosion and plate tectonics, but the ground-down Vredefort structure in South Africa can still be recognised for what it is—at least it can if you look down from orbit and see the Earth as the Earth sees the Moon.

● ○ ◐

As the Earth saw the Moon.

Now it has seen it differently. It has seen it in close-up from orbit and from the surface—most intimately in the many thousands of exposures captured by the Hasselblads of the Apollo astronauts.

The surface is not, in absolute terms, bright. In its nature tarmac dark, reflecting only about 12% of the sunlight that hits it. When evenly, brilliantly lit, though, it can but look bright.

It is not quite monochrome. Its colours are faint, and they are almost all variations on shades of grey, but there is at least a hint of pigment: something reddish, or something bluish. The slopes, if there are slopes to be seen, have a different tone from the plains below. Those that rise most steeply show the most changed texture; nothing loose clings to them, and they are but rock.

There is something complete about those rising forms. They are unfurrowed; nothing has rasped at them, or for the most part, cut through them. They never double back on themselves. Their subdued relief is not without variation, but its range of expression is restricted. The long heights lack any pattern save that of slow curves.

They cast shadows, though. In the photographs, the shadowed sides of the Moon's distant hills and rilles look as black as the sky. In reality, they are not—they are lit by the Sun as reflected from nearby sunlit surfaces. The daytime shadows of the Moon are lit by moonshine. The night is lit by earthshine.

As yet, no one has seen that nocturnal illumination from the surface. The Apollo astronauts saw it from orbit, though. Ken Mattingly, who flew Apollo 16's command module, *Casper*, told the author Andrew Chaikin that the experience was "like flying over snow-covered terrain on the Earth with a bright moon and totally clear skies. You get this magic terrain—you can see relief. But it has this sameness, this uniformity in color. . . . [I]t's just like that—except you can see more detail, because Earthshine is so much brighter than moonshine."

The astronauts on the surface found distance difficult to read. The idea that, on a smaller world, the horizon is closer is easily understood. In practice it is hard to gauge. Evolution has given humans a strong intuition for how far their eyes can see when they stand on a flat plain. The Apollo astronauts never quite got the hang of overriding that intuition, endlessly thinking features in the landscape were farther away than the maps said they were.

It is not that the landscape in the photographs lacks telling detail; it has features and objects in it, rocks and boulders which sit proud of the rubble and grit, distinct and individual. But the details mostly make no sense. A few do: a boulder at the bottom of a slope, having rolled or slid down when dislodged from higher up by a Moonquake, is easily read. Sometimes there is a track down the slope above it that shows its trajectory. In general, though, there is no process to be inferred here. The landscape may have features that move one into another, slopes that become plains, ridges that roll back, but they do not have stories in the way a river's valley does. It is, after all, just the work of impacts. The Moon's timescape has no flow; just punctuation.

The exception to the landscape's resistance to measurement lay in the Earth-objects that many of the photographs capture. Looking back at their landing sites, the astronauts knew how far away they were. When *Intrepid*, the lunar module for Apollo 12, landed a few hundred metres from Surveyor 4, an extraordinary feat of precision navigation, its crew guessed the distance to the other spacecraft easily. Human detail was legible in a way the lunar landscape was not.

The rest of the rocks are just sitting where they were left. There is a sense of things having been abandoned. It is the opposite of a pause, not a stasis that interrupts a process but a stasis that is the norm.

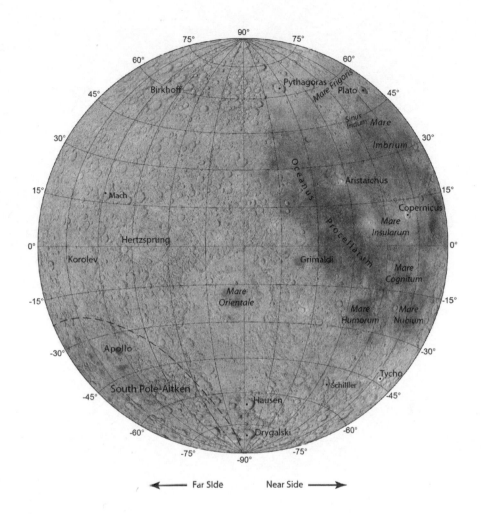

THE MOON'S WESTERN HEMISPHERE

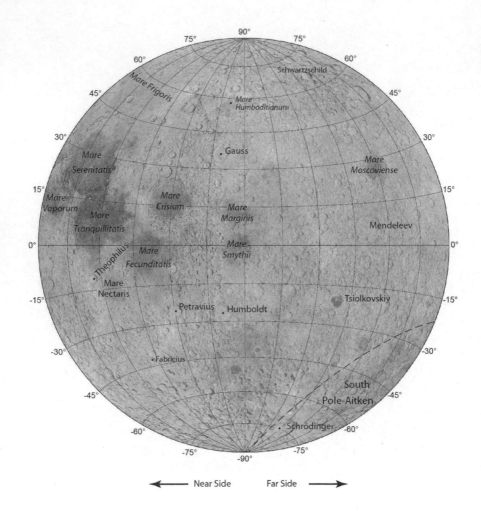

90°
75°   75°
60°   60°
Schwartzschild
45°   45°
Mare Frigoris   Mare
Humbodltianum
30°   30°

Gauss

Mare   Mare
Serenitatis   Moscoviense
15°   15°
Mare   Mare
Vaporum   Crisium
Mare   Mare
Tranquillitatis   Marginis
0°   0°
Theophilus   Mare   Mare
Fecunditatis   Smythii   Mendeleev
Mare
Nectaris   Tsiolkovskiy
-15°   -15°
Petravius   Humboldt
-30°   -30°
Fabricius
South
Pole-Aitken
-45°   -45°
-60°   -60°
Schrödinger
-75°   -75°
-90°

← Near Side   Far Side →

THE MOON'S EASTERN HEMISPHERE

ITS ORBIT

FROM THE EARTH TO THE MIDDLE OF THE NEARSIDE—ITS CLOSEST point and thus, because its mappers are of an Earthly bent, home to its prime meridian—is a distance about 60 times the radius of the Earth. If the Earth were the size of your head, the Moon would be a smallish apple on the other side of your living room.

The exact distance can be as much as 398,600km or as little as 348,400km, because the Moon's orbit is an ellipse, not a circle. As a result, its apparent size, as seen from the Earth, also varies: bigger when closer. When it is at its closest—a condition called perigee—the Moon can look as much as 14% larger than when it is at its farthest (apogee). Full Moons which coincide with perigee have come to be known as supermoons and can be up to 30% brighter than other full Moons.

The plane of its orbit is not the same as the plane in which the Earth orbits the Sun, which is called the ecliptic; if the two planes were the same, every new Moon would eclipse the Sun as the three bodies lined up precisely. But the Moon's orbit does cross the ecliptic—the points where it does so are called the nodes— and if the Moon is full or new when it reaches one of those nodes, observers on Earth will see an eclipse. If the Moon is new, it will be an eclipse of the Sun. If the Moon is full, the Earth's shadow creates a lunar eclipse.

Though the Moon's orbital plane is not the same as the ecliptic, the angle between the two is small. This means that the Moon's path through the Earth's sky, like that of the Sun, changes with the seasons. In winter, when the Earth is leaning away from the Sun, both the Sun and the new Moon stay lower in the sky than they do in summer; the full Moon, opposite the Sun, rises higher and stays up longer. Above the Arctic Circle the winter Moon can stay up for more than 24 hours at a time, just as the summer Sun can. A scientist in Spitsbergen once told me that the midnight Sun is not as special to him as the noontime full Moon.

As the Moon travels round the Earth, it moves with respect to the fixed stars behind it. Because its path is quite like the Sun's, its circuit mostly passes through the signs of the zodiac (which are defined by the ecliptic). The Moon takes 27 days, 7 hours and 43 minutes to complete this circuit, a period of time called a sidereal month (from the Latin for "starry", as in "Sidereus Nuncius").

The 12 constellations of the zodiac are not ideal for keeping track of the orbit of the Moon. Because it does not orbit exactly in the ecliptic it strays north and south of them a little; and 12 signs is not a handy way of dividing up a 27.3-day circuit. Chinese astrology instead divides the Moon's path into 28 mansions.

It is this movement against the background of the fixed stars that explains why the Moon rises a bit later every day. On the day that I am writing this, the Moon was in the mansion of the Net when it rose, which is defined by the star Epsilon Tauri; tomorrow it will have moved on to the mansion of the Turtle Beak, defined by Meissa, the star at the tip of Orion's sword, which will rise about 50 minutes later.

Just as the Earth must make more than one revolution to catch up with the moving Moon, so, because the Earth is moving round the Sun, the Moon must make more than one orbit round the Earth to get from one full Moon to the next. Imagine the Moon at the end of the minute hand of a clock centred on the Earth, and the Earth on the hour hand of a clock centred on the Sun. When both hands point the same way—let's say at the nine—the Moon is full. An hour later the Moon will have made a complete circuit and be pointing at the nine again. But the Earth-hand, as it were, will have moved on to

the ten with respect to the Sun. So it will be another five minutes before the hands line up again, with the Earth between Sun and Moon
and the Moon back to its fullness. This longer period—29 days, 12
hours and 44 minutes—is called the synodic month.

The synodic month is also the time that it takes the Moon to turn
on its axis; that is why the nearside stays near and the farside, far. This
is not a coincidence; it is an effect of the Earth's gravity known as tidal
locking. But the lock does not hold as tight as it might. There is some
wriggle room. The Moon rotates at a constant speed. But the speed
at which it moves in its orbit changes according to the laws Kepler
laid down for elliptical orbits: at apogee it moves a touch slower;
at perigee, a touch faster. So, near perigee, the Earth sees a little
more of the Moon's western hemisphere—because the Moon has
got ahead, as it were—and at apogee, as it slows down, a little more
of the eastern one. This "libration"—the effect that so interested
William Gilbert, the first Moon mapper—means that if you watch the
Moon closely enough for long enough, you will, over time, see rather
more than half of it. Still, more than 40% of its surface is never seen
from the Earth.

Astronauts on the International Space Station are at times closer
to the Moon than all the rest of humankind. But just by a tenth of 1%.
Only 24 men have, to date, gone closer than that.

- III -

APOLLO

I THINK, THOUGH I CANNOT SAY FOR SURE, THAT GENE SHOE-maker was the first human being to know that he could, that he should and that he very well might walk on the Moon.

It was 1948, the summer after his graduation, and he was doing fieldwork in Arizona, camping under the stars. His alumni newsletter included a story about Caltech researchers experimenting with captured German V-2 rockets a few hundred kilometres to the east of him, at White Sands in New Mexico. With that news to hand, he saw the double nature of the Moon as no one had ever done before; he saw a place, sickle sharp in the high, clear sky, where a man like him might walk and work.

"We're going to explore space," he later remembered thinking, "and I want to be part of it! The Moon is made of rock so geologists are the logical ones to go there—me, for example!" It was to make himself the logical choice that he became the world's greatest expert on cratering; it was also the reason he created the US Geological Survey's astrogeology programme.

Two decades later, on a bright Florida morning, a cylinder of thin ice the size of a grain silo hung suspended tens of metres above the ground. The frost had started forming in the middle of the floodlit night, when the technicians at Cape Kennedy had started to fill the great tank at the top of the Saturn V's first stage with liquid oxygen— more than a million litres of it, at a temperature of minus 183°C. The wall of the tank and the skin of the rocket were one and the same, so water vapour from the humid Atlantic air had immediately started to freeze to the painfully cold metal.

As the oxygen was pumped in, some of it boiled off; vents at the top of the tank let the vapour out so that pressure within would not get too high. At 09:30, the vents were closed. Helium was pumped into the small space at the top of the tank. The pressure started to rise.

Below the oxygen tank was a slightly smaller tank filled with highly refined kerosene. Below that, arranged like the dots on the five face of a die, were the F-1 engines, exquisitely engineered, cunningly contrived, ludicrously powerful.

Two minutes after the vents were sealed, a valve at the bottom of the upper tank opened, and oxygen began to flow down into the F-1s. It took two different routes. Some of it went into gas generators which were linked to turbines which drove pumps. In the generators, it was mixed with kerosene and sparked alight. There was too much kerosene for the not-yet-full flow of oxygen to consume it all; the hot exhaust that the generators passed to the turbines was dirty black with part-burnt fuel. That didn't stop it from spinning them up and bringing the engines' pumps to life.

The rest of the oxygen went into the combustion chambers proper. There it met the kerosene-rich exhaust coming out of the turbines, and the mixture was set alight all over again. Black smoke began to billow from the bottom of the F-1s' nozzles. The rocket began to shake. The pumps increased the flow of fuel and oxygen down into the fires below.

A carefully choreographed dance of temperature and energy was now under way. The turbopumps used energy from the fuel burned in the generators to get ever more fuel into the combustion chambers, but they sent it there by way of a spiralling detour through tubes wrapped around the engines' nozzles. This cooled the nozzles, which otherwise could not have borne the heat they were subjected to. It also warmed the fuel, which thus burned even better when, at last, it reached the combustion chamber. The fuel was also the lubricant for many of the engines' moving parts—and the soot produced early on gave the lower section of the nozzle more protection from the heat of the growing flame within.

The pumps spun harder; the dance sped up. Five seconds after ignition, the fuel valves were fully open, and within a second or so the engines were close to full thrust. The central engine came to full power first, then the four outer ones. The fuel mix was now richer in oxygen, the burn cleaner and less sooty, more powerful. For a second or two after the last engine came up, the rocket was held down by mighty clamps. Then it was released.

All the weight of the mighty first stage, of the propellant-filled second and third stages and of the *Eagle*, the *Columbia* and its service module right at the top—almost 3,000 tonnes in all—now rested on the engines. They shouldered their burden and began to lift. The five arms from the tower that steadied and fed the rocket swung back. The shell of ice that had clung to the supercool metal fell in shattered sheets into the inferno below.

The fires on which it rose were not the fire that leaps or licks or plays, the fire of brasier or boiler. They were the focused fire of the metalworker's torch, given life at a scale to cut worlds apart or weld them together. The temperature in the chambers was over 3,000°C. The pressure was over 60 atmospheres. And still the pumps, their turbines spinning 90 times a second, were powerful enough to cram more and more oxygen and fuel into the inferno. The flames slammed into the fire pits below at six times the speed of sound. For a couple of minutes, the five F-1s generated almost 60 gigawatts of power. That is equivalent to the typical output of all Britain's electric-power plants put together.

It took ten seconds for the rocket to clear the tower. It took a further ten seconds for the roar of its engines, louder than any noise humans had made before, to reach the VIP stands almost 6km away. Sixty ambassadors, half of Congress and about a quarter of America's governors, watching with awe, shaken by "a sound that became your body", as the artist Robert Rauschenberg put it.

The roar lasted less than three minutes. But by the time the F-1s fell silent, the rocket was travelling at over 8,000kph and was 600km from Cape Kennedy. Apollo 11 was on its way to the Moon.

Gene Shoemaker was not aboard. He had not been able to become an astronaut; Addison's disease had put paid to the Moonshaker's dreams of being a Moonwalker. The astrogeology programme that he had created, though, shaped where the astronauts would go on the Moon and what they would do there. It had trained all three of the men flying off over the Atlantic that morning, and all their successors, too.

Shoemaker could have had an honoured seat in the VIP stands. But he and his wife, Carolyn, were rafting down the Colorado River.

● ○ ◉

THE V-2S THAT SHOEMAKER HAD READ ABOUT IN HIS ALUMNI newsletter started as inaccurate terror weapons. They killed thousands in Belgium and Britain. The death toll among the concentration camp slaves in the factories where they were built was even higher.

The men who designed and built them had come, earlier than Shoemaker, to believe that space travel was a real possibility. Many of them had been members of the Verein für Raumschiffahrt (VfR), an association of German rocketry enthusiasts inspired by the ideas of Hermann Oberth, as set forth in his book "Die Rakete zu den Planetenräumen" (1923; "The Rocket into Planetary Space"). Some of them went on to work on the Saturn Vs that took Americans to the Moon.

Oberth, a physicist and engineer originally from Transylvania, was one of a trio of early visionaries who believed both in the goal of space travel and in a specific technology that would deliver it: the liquid-fuelled rocket. Rockets of the sort first developed in China, used for both celebration and warfare across Eurasia since the Middle Ages, burn a solid fuel, such as gunpowder, to produce hot, expanding vapours. Expelled backwards, the gases drive the projectile forwards.

As a use of gunpowder, such rockets were rarely preferred to guns themselves: the hot, expanding gases provided kinetic energy more effectively when confined by a metal barrel so as to expel a small projectile in a reasonably predictable direction. But the preference was not universal. The Kingdom of Mysore in India pioneered the craft of making the rockets themselves from metal.

The British, having found themselves on the receiving end of this weaponry, liked the rockets' rate of fire and portability, when compared to cannon. They applied the new metalworking skills of the industrial revolution to improving the technology; their rockets' red glare over Fort McHenry, Maryland, was soon commemorated in America's national anthem for its frightening drama, if also for its lack of effect. In 1861, four years before Jules Verne used a gun to send his travellers to the Moon, a Scottish astronomer called William Leitch suggested that rockets might fulfil the same office.

Rockets, yes. Solid rockets, no. The gunpowder used in rocketry is a mixture of fuel—sulphur and charcoal—and oxidant—potassium nitrate, also known as saltpetre. It is the fact that the fuel reacts with an oxidant readily to hand, rather than needing air, that makes gunpowder a useful explosive. But the rate at which the explosion can take place is limited by the speed that the explosion moves through the gunpowder. In a liquid-fuelled rocket, the fuel and the oxidant can burn as quickly as you can pump them together. It was a Russian physicist, Konstantin Tsiolkovsky, who first showed that a rocket which used liquid hydrogen as a fuel and liquid oxygen as an oxidant could, if built of multiple stages, attain a speed of 8km/s, enough in principle to put it in orbit round the Earth.

This was not, to Tsiolkovsky, a mere fact of engineering. To attain orbit was the most important thing any machine could do. Like most speculative thinkers of his day, Tsiolkovsky, one of a group of Russian thinkers known as "Cosmists", was seized by the importance of evolution. But he did not look to space, as James Nasmyth did, to understand the past evolution of planets, but to permit the future evolution of people. He saw space-based evolution in terms of the perfection of humanity, the realm in which life could become carefree and immortal, illuminated by endless cosmic energy.

Oberth built on Tsiolkovsky's research, as did Robert Goddard in America. Both men were, like their Russian inspiration, seized by the possibility that humans could travel beyond the Earth, though they tended not to express themselves on the matter as mystically as Tsiolkovsky did. They also knew they faced daunting technological challenges, such as storing and pumping liquid oxygen and crafting combustion chambers and exhaust nozzles which could withstand uncommon pressures and temperatures. Solving these would require years or decades of expensive research. There were applications beyond, or perhaps before, spaceflight: rockets might be a way of delivering emergency supplies, and maybe even first-class mail, to otherwise hard-to-reach places—the delivery drones of their day. Oberth gave lectures on such possibilities. But, in his memoirs, Willy Ley, vice president of the VfR, recalled a conversation with the older man after a lecture in the late 1920s:

> "Do you think, Herr Professor, that there will be a need for rockets carrying a load of mail over five hundred kilometers?"
>
> Oberth looked at me with the smile which old-fashioned pedagogues reserve for people whom they call "my dear young friend" and said after a while: "There will be need for rockets which carry a thousand pounds of dynamite."

And this was, indeed, the killer app. Goddard's project was not obviously military; he attracted money from various sources, including the Smithsonian Institution and the Guggenheim family. Goddard was introduced to the Guggenheims by Charles Lindbergh, the aviator and fascist sympathiser who had become fascinated by talk of a rocket that could reach the Moon.* But during the 1930s the armed forces were his biggest sponsor. At the same time, with the rise of the Nazi Party, the VfR saw many of its leading lights transfer themselves to the Wehrmacht. Rocketry, which had hardly contributed at all to the carnage of the First World War, was of such little account that the Treaty of Versailles had neglected to forbid or even mention it, meaning it was an area of weapons development where Germany was unconstrained.

The VfR's counterpart across the North Sea, the British Interplanetary Society (BIS), founded in 1933, suffered an opposite regulatory fate: Britain's Explosives Act of 1875 meant it could not build rockets at all. It thus found itself taking rather more of an armchair approach to the subject than the VfR or the American Rocket Society, founded as the American Interplanetary Society in 1930. The first time the BIS's members got a true experience of what a rocket could do was when a V-2 hit near a London pub where some of them were drinking in late 1944. They raised their glasses to their comrades overseas.

One of those drinkers was a young Royal Air Force officer called Arthur C. Clarke.

● ○ ◐

IT IS COMMON NOW TO THINK THAT SCIENCE FICTION IS ABOUT the future, but this was not always the case. The 19th-century works most often discussed as science fiction—Mary Shelley's "Franken-stein", Verne's "From the Earth to the Moon", H. G. Wells's "The War of the Worlds"—do not take place in the future. They take place in their own presents, but with something new and unsettling added: a creature created by man, a cannon that can break the boundaries of the Earthly realm, an alien imperialism capable of doing to London

---

* In his novella "US" Howard Waldrop touchingly imagines three futures for Lind-bergh's son, Charles junior, kidnapped and killed in 1932. One of them is to become the first man on the Moon.

what London did to others.* The disruptive novelties with which the books challenged their readers relied for their power on their eruption into the present. Verne's only novel explicitly about the future, "Paris au XXe Siècle" (1863; "Paris in the 20th Century") was not published in his lifetime.

As the long 19th century drew to its end, though, a new notion of the future started to come into its own. This modern future differed from the way the future had been thought about before. For one thing, there was more of it. Religion had, until fairly recently, held a firm grip on both time future and time past. Geology's Deep Time, and Darwin's, had loosened its backwards grip; physics was breaking its grip entirely as it opened up even greater past depths—and looked forward to a deep future, as well. Christianity had said the world would come to an end; thermodynamics promised the same thing. But the amounts of energy that had been discovered in the atomic nucleus promised relief. By 1906 Frederick Soddy was able to talk of a universe lasting effectively forever thanks to the near infinite powers of radioactivity. Whichever way science looked, the depth of field was rushing away from the present in a sort of doubled-up cosmic dolly zoom.

This expansion in time had a correlate: a shrinking of the world in space. The smallness of the Earth is a long-standing trope. People imagined the Earth as small long before "Earthrise" showed it as such—Thomas Digges, England's first Copernican, imagined how small it might look from space even before Galileo did. But the smallness was felt ever more strongly as the century of the railway and the steamship drew to a close. America, it was claimed, had lost its frontier. There were no uncharted oceans in which to locate that archipelago of deep thought and high jinks where tales of the fantastic used to live, less and less terra incognita in continental interiors for lost worlds to be lost in. The poles themselves were about to be

---

* This is one of the reasons why it is footling to read texts like "From the Earth to the Moon" as either successful/accurate (three-man crew, launched from Florida) or unsuccessful/inaccurate (a gun, for God's sake?) predictions, judged against the realities of 1969.

conquered; soon men would fly through the air in powered machines, not just balloons.

In 1969 Norman Mailer would tellingly describe the way in which landing on the Moon changed the world as feeling like a sort of geometrical inversion, a pocket being turned inside out. The same is true for time and space at the beginning of that century. Space closed down; time opened up: what could no longer be contained in the former was pocketed in the latter. In cinemas, lamps and shutters turned physical lengths of film into stories told in time. In Einstein's physics, time and space were traded off relative to one another, with only the speed of light remaining constant. Indeed, speed was to become a frontier in itself, as the century saw record after record fall to the motor car, the aeroplane, the rocket. "Speed," the writer Aldous Huxley would remark, "provides the one genuinely modern pleasure."

The changing geometries of time and space made it almost natural that fictions or satires which until recently would have been set elsewhere were instead set elsewhen. They opened a new geography for the mind. And it was not just for fancy. As H. G. Wells put it in 1902, in a talk entitled "The Discovery of the Future":

> I believe quite firmly that an inductive knowledge of a great number of things in the future is becoming a human possibility. I believe that the time is drawing near when it will be possible to suggest a systematic exploration of the future.

Wells intended his work, both fiction and non-fiction, to be part of that exploration. Countless imitators and followers were to claim the same mantle, and the basis on which they did so was to reshape the way the future was thought of. The future became scientific, and science fictional, in three related ways. Science was, in itself, predictive, and its remit was steadily spreading through society. Science felt increasingly tied to technology, and technology was coming to define both modernity and futurity—as manifested, for example, by that fixation on speed. And to socialists like Wells, science also shaped what the future should be, defining both the proper human relation to nature, and the proper human attitude to social control—that of the white-coated technocrat. Through such men, the scientific facts of the world would impose harmony on society.

The younger you were, the more of this thrilling future there was. Science fiction provided such readers with both thrills and, as it came into its own, reassurance. As they grew up with it, it told them that if they understood science, they could understand the often-baffling world as it would be and should be. They could look forward to societies that made sense because they had to, a world in which things had their rightful place, and the rightful place for smart people was an honoured one.

As publishing became a business of creating and maintaining identifiable genres, this future became for science fiction what the West was for the Western—a settled substrate out of which writers could carve distinctive landscapes as they played with the tropes the readers had paid for. The foremost of those tropes was the spacefaring rocket, as central to science fiction as the horse to the Western. It embodied the underlying transformation of distance into time; the worlds it visited were, by definition, the worlds of the future.

The pioneers of liquid-fuelled rockets, inspired by fictions from the 19th century, set about creating real machines in the context of this 20th-century futurity. Oberth was said to have reread "From the Earth to the Moon" so many times as a child that he knew it almost by heart. As an adult, he was delighted to act as an adviser on Fritz Lang's "Frau im Mond" (1932; "The Woman in the Moon"), advertised as "the first utopian film based on scientific fact". The film provided a new concern for realism to the cinema of rocketry. Lang showed that a rocket to the Moon would have to be built of several stages: only if empty fuel and oxidant tanks were discarded would the spacecraft proper be light enough to get into orbit. He also imagined a mission to the Moon which resulted in something like settlement, rather than just an adventure in which the outcomes were either death or return, and in which the conflict that drove the plot was entirely between humans, not aliens: his Moon was uninhabited. In a lasting contribution to the drama of real spaceflight, he had a character count backwards to zero to increase the tension as the launch approached, a custom which the German rocketeers adopted from then on.* Most important of all was the innovation right there in the title: his travellers included a woman.

Goddard, for his part, was first seized by the possibility of space-flight shortly after he read H. G. Wells's "The War of the Worlds" (1898) and Garrett P. Serviss's unauthorised sequel, "Edison's Conquest of Mars" (1898), in which the great American inventor and his comrades create a vast, very expensive fleet of spacecraft with which to take the fight back to the Martians. Goddard was, he would later tell people, pruning a cherry tree when a reverie came over him; he imagined a rocket ship ready for Mars on the meadow beyond the garden.† The people who followed in the slipstream of Oberth and Goddard—members of the VfR, the BIS and the American Rocket Society—were science fictioneers all but to the man. Robert Heinlein joined the American Rocket Society in 1932.

The Second World War was crucial to the development of both liquid-fuelled rockets and science fiction. At the V-2 development site in Peenemünde, on Germany's Baltic coast, Wernher von Braun and his colleagues achieved advances in the state of their art similar in magnitude to those made at Los Alamos, where the atomic bomb was created, and the MIT Rad Lab, where radar technology was developed. The peroxide-powered turbopumps and temperature-resistant combustion chambers of the V-2 allowed its engine, only 1.5m long, to produce more power than a battleship. The rockets could achieve altitudes of 100km or more.

At war's end, America, Britain and the Soviet Union all rushed to get their hands on the hardware and the people who had designed

---

* It is telling that the countdown came from cinema, a medium in which time is encoded in a form that can move equally easily forwards and backwards.

† When the beloved tree was brought down by a storm in 1938, Goddard noted sadly in his diary that he would "have to go on alone".

it. Hence, three years later, the experiments at White Sands of which Shoemaker read at his campsite in the pines.

Rocketry was only one of the mainstays of the science-fictional future. Superweapons were another. Devices capable of projecting genocidal slaughter had been part of the futures imagined in pulp magazines and elsewhere since "The War of the Worlds"; how else would a fictionalized Thomas Edison have conquered Mars? Such weapons were not only, nor even mostly, used on aliens; they were frequently and uncritically unleashed on other humans, normally of races other than white. John W. Campbell junior, who ran the world's leading science fiction magazine, *Astounding*, was sure that the war would see such superweapons realised through the use of atomic power. In 1943 he fed one of his authors the technical know-how with which to write a story about atomic weapons accurate enough to earn the magazine a visit from the FBI, which had been tipped off by *Astounding* readers who were working on atomic weapons for real in Los Alamos. After Hiroshima, that FBI visit became a much-touted badge of validity for science fiction's prophetic pretensions.

The potential for combining the power of the rocket and the power of the atom was obvious, and unnerving, to Campbell and the writers in his circle not least because, though America had the lead in nuclear weapons, Germany had demonstrated that other countries could have the lead in rocketry. In the months after the destruction of Hiroshima and Nagasaki, Heinlein frantically tried to alert former comrades in the Navy to the risks; in May 1947 he put the idea into print in "Rocket Ship Galileo". The book's teenage heroes find Nazis who have taken refuge on the Moon planning to nuke America.

The same month's issue of *Air Trails and Science Frontiers*, a non-fiction magazine Campbell was also editing, published "Fortress in the Sky", an account of the idea on a missile base on the Moon by another of Campbell's *Astounding* writers, L. Ron Hubbard. In 1948 *Colliers* published the altogether more disturbing "Rocket Blitz from the Moon" by Robert Richardson, an astronomer who also wrote fiction for Campbell. It was chillingly illustrated with pictures of sleek

missiles launching from craters on the Moon and of Queens and
Manhattan under multiple mushroom clouds, all from the brush of
Chesley Bonestell, a friend and collaborator of Richardson. The destruction wrought on the five boroughs deserved an artist used to the
planetary scale; indeed, the pictures serve as companion pieces to one
that Bonestell had painted the year before showing the aftermath of a
meteor strike on New York.

As the Moonpeople's self-destruction through nuclear war in
"Rocket Ship Galileo" made clear, the idea of the Moon as a source
of mass destruction resonated with its long-standing fictional deploy-
ment as a landscape of deathliness, a place of extinction sometimes
dotted with ruins. In "Around the Moon" Michel Arden imagines the
desolate plain beneath him as a giant ossuary. At the climax of C.
L. Moore's peculiar and powerful planetary romance "Lost Paradise"
(1930), her hero, transported in time, sees the gods of the Moon kill
their world by tearing off its atmosphere. The lunar landscape became
a foreshadowing of what the Earth would look life if nuclear weapons
were used again, whether those weapons came from the Moon or not.
In an intriguing and disturbing reversal of this idea, in the 1950s the
US Air Force studied the possibility of launching a nuclear weapon to
the Moon in order to see what sort of crater it made—and to make
clear to everyone else that the power to reach beyond the world and to
end it were in the same hands. Wiser heads prevailed.

The possibility of moonbases as launch sites was not confined to
science fiction. Aerospace companies approached by the US Air Force
to design such a base in the 1950s included missile silos in their plans
even though they were not explicitly asked to—it just seemed like the
thing to do. But Earth-based nuclear weapons which popped up into
space only long enough to fall back down half a world away were the
more practical way to go.

The new emphasis science fiction put on the Moon was not limited
to matters military: indeed, "Destination Moon" (1950), a film based
to some extent on "Rocket Ship Galileo", had the Nazi base removed.
Heinlein wanted to make a Moon film that was more realistic, less

"Buck Rogers"—part of his plan to make Americans believe in space-flight. Writing the screenplay brought Heinlein joy—he got to work with Chesley Bonestell, who painted the background mattes for the film—and worry—the eventual producer, George Pal, apparently toyed with turning at least some of the piece into a musical. He brought on James O'Hanlon, who would go on to write "Calamity Jane" (1953), to punch up the script.*

No music or dance made it into the final cut. I rather miss them; the low gravity of the Moon might lend itself to dancing, and the rather tedious "Destination Moon" that was finally made could do with a lift, in every sense. But Heinlein wanted both his career and the prospect of near-term missions to the Moon to be taken more seriously. He was selling Moonstuff to as wide a range of markets as he could: "Nothing Ever Happens on the Moon" went to *Boy's Life*, the magazine of the Boy Scouts of America; "The Long Watch" went to *American Legion* magazine; "Space Jockey" went to the *Saturday Evening Post*, America's biggest circulation "slick" magazine (as opposed to "pulps" like *Astounding*); "The Man Who Sold the Moon" was to be the never-before-published title story in a new hardback collection of his work. Learning that girls as well as boys were reading his stories, he wrote a particularly nice one just for them, "The Menace from Earth". His aim was to write science fiction that would appeal beyond the narrow boundaries of the genre—which meant everyday stories of scouting, career choices and dating that happened to happen on the Moon.

Clarke, the other science fiction writer most interested in working to make spaceflight real, was doing something similar with stories that were intriguing and technically adroit but studiedly unmelodramatic. He was also writing successful non-fiction works such as "The Exploration of Space" (1959). Admittedly, his novel "Earthlight" (1955) does have a spectacular set-piece battle, but it is seen

---

* Strangely, the archives of Allan Grant, a photographer assigned to a story on "Destination Moon" for *Time* magazine, contain a set of photographs of an apparently filmed dance piece taking place on the film's set. What the performance was remains unknown.

from the point of view of an astronomer and accountant who happen to be nearby, not from that of the belligerents.* Most of the book is about observatory life, which apart from taking place on the Moon a couple of centuries hence is much the same as it would have been on Earth in the 1950s—complete with photographic plates used to record images and "computers" being young women with calculating machines.

Clarke's "A Fall of Moondust" (1961) is the cleverly crafted story of a lunar "shipwreck". It drew on the idea put forth by Thomas Gold, a brilliant and iconoclastic physicist, that some, at least, of the Moon's dust might be so fine as to be effectively liquid. This turned out not to be true, and Gold's insistence that it might be made him something of a bête noire for the astrogeologists, but it made for a good story. Clarke marries his expertise in lunar science and his grasp of space engineering to a puzzle plot in which the crew of the ship that has sunk in the dust and their would-be rescuers have to come up with solutions to an endlessly escalating string of problems. The combination of a tightly paced plot, well-handled pedagogy, an exotic frontier setting—the comparison to stories of the American West is archly explicit—and relatable characters—the ship is a tourist vessel, full of passengers to whom the Moon is almost as strange as it is to the reader—may have been what made it, in the winter of 1961, the first piece of science fiction ever to be run as one of the "condensed novels" in *Reader's Digest*, something which garnered it a far greater readership than a science fiction novel could normally expect.

It seems more likely, though, that *Reader's Digest*'s decision may have rested on something else. A respectably written book with "Moon" in the title had a new salience that fall. Six months before, President Kennedy had dramatically committed real-world America to the science-fictional task of putting a man on the Moon.

● ○ ●

---

* OK, the accountant is a spy, but that hardly matters.

THROUGHOUT HIS ADULT LIFE HARRY TRUMAN CARRIED A PIECE of science fiction in his wallet with him—six strange, prophetic stanzas from Alfred, Lord Tennyson's "Locksley Hall" (1835). They speak of a future age of superweaponry and the "Parliament of Man, the Federation of the World" that the sheer power of those weapons force into being. They are lines which were echoed through much of the superweapon fiction of the early 20th century. Simon Newcomb's "His Wisdom, the Defender" (1900), an influential prediction of the destructive power that aerial bombardment would have, quotes them directly; H. G. Wells's "The World Set Free" (1914), a speculation about atomic power and weaponry that was required reading on the Manhattan Project, echoes their sense closely.

Truman recited from the poem on the way to his meeting in Potsdam with Joseph Stalin and Winston Churchill. It is hard to imagine its lines were far from his mind when he ordered the use of the first atomic bomb—or when, two days after the heart of Hiroshima was destroyed, he signed the charter of the United Nations.

John F. Kennedy was not a science fiction reader; thrillers were more his thing. He had no Tennyson in his wallet and no interest in rocketry. But what Truman did for the superweapon, Kennedy did for that matching pillar of science fiction, the Moonrocket.

His motivation was geopolitical. The Soviet Union did not get quite as good a crop of V-2 scientists from Germany as America did, but it got a lot of technical know-how. And it had better people geared up to use it (Robert Goddard died in 1945). The cosmist philosophy of Tsiolokovsky and his coterie of peers (they included Vladimir Vernadsky, whose ideas about the biosphere as a self-regulating Sun-driven machine foreshadowed, in some ways, Lovelock's Gaia) had, with a touch of *mutatis mutandis*, made the transition to communism in reasonably good shape, providing as it did a heady mix of materialism and destiny. The advent of nuclear weapons meant that the Soviet Union would need rockets for less spiritual needs—and big ones, too, because its nukes were, early on, large and heavy. And it had a truly inspired engineer, Sergei Korolev, who could put the resources Stalin provided him with to good use. The result was the R-7—an intercontinental ballistic missile capable of putting a satellite into orbit. On October 4th 1957 it did so.

Americans, believing themselves to have been in the van of technological progress, were shocked. They were also afraid. Sputnik made it clear that the USSR could put any place in the world in jeopardy. In the election campaign three years later, Kennedy stoked that fear, emphasizing the idea of a "missile gap" between the two superpowers. There was no such gap; the R-7 was never deployed operationally. America's U-2 spy planes gave Dwight Eisenhower's administration a strong indication of this; by late 1960 America's first CORONA spy satellites were confirming it. But that was too late to change the election results—and could not be talked about anyway, because the satellites' mere existence was beyond top secret.

The team under von Braun at the Army's Redstone Arsenal, outside Huntsville, Alabama, put America's first satellite up a few months after Sputnik, in January 1958, on a pretty direct descendant of the V-2. They would have been quite capable of launching one earlier. But the Army was meant to build only medium-range missiles. The job of making ICBMs had been assigned to the Air Force, which was not overwhelmingly seized by the possibility: rockets don't have pilots. The satellite programme, meanwhile, was under the control of the Navy.

Though Eisenhower was cagey about spending on space flight, later in 1958 he created NASA and approved Project Mercury, which would eventually use an Air Force ICBM, the Atlas, to put Americans into space. In 1960 he even reluctantly approved the development of a rocket specifically for the launching of spacecraft, rather than warheads—a so-called super booster that would be developed as a successor to von Braun's Jupiter rockets and named Saturn. This was the project that created the F-1 engine. But when it came to talk of grand ambitions, Eisenhower was completely dismissive: NASA's first administrator, Keith Glennan, remembered him saying that "he couldn't care less whether a man ever reaches the moon". His last budget, finalized in January 1961, contained no money for any human spaceflight activities beyond Project Mercury.

Kennedy and his advisers were not, to begin with, noticeably more enthusiastic. He did not commit to anything beyond Project Mercury, from which his advisers were encouraging him to distance himself; it was unlikely to get a man into space before the Soviets, and being too closely associated with a failure would be an unforced political error. On April 12th Yuri Gagarin duly went into orbit on a Vostok launcher. The achievement was not unanticipated, but the level of excitement seen around the world was. Kennedy may not have read science fiction, but he certainly read newspapers; he decided that America needed to take the lead. On April 20, after America had suffered another humiliation, this time in Cuba's Bay of Pigs, Kennedy asked his vice president, Lyndon Johnson, to let him know "at the earliest possible moment" if there was "a space program which promises dramatic results in which we could win?"

The idea was not simply to look good or to make up for embarrassment. There was more at stake than that. The atom bomb and the spaceship were the most dramatic examples of what was being treated as a new era of technology—an era of worldwide systems and godlike powers, born from and growing up in conflict. The idea that a centrally planned economy might offer the best way of harnessing this new power was not implausible. The USSR might be better at this new task of the future because it was better suited to the future in general—and thus a better model for the countries of the global South then coming into their independent own. After Apollo, and after the fall of the Soviet Union, it was for decades hard to remember a time when the *Dallas News* could muse as to whether there might be "some advantages of tight, totalitarian control" when it came to technology. But the idea that the American system might be too soft and too consumerist to bring forth and marshal the world-changing technologies of tomorrow had genuine currency.

The answer that the experts provided when asked what America could do to prove this wrong was to double down: choose a space project so challenging that a whole new level of technology would be needed, thus rendering the Soviets' first-mover advantage moot. A

Moon landing was just such a project. As von Braun, the man in charge of super boosters, self-servingly but accurately pointed out to Johnson, it would require rockets ten times more capable than the current state of the art. Yet, with the F-1 in development, it could be delivered, he and others assured the White House, by 1968 or so—which is to say, before the end of Kennedy's second term in office.

At various times Kennedy wondered out loud whether pushing a breakthrough in desalination might not be as striking and far more useful. But he started to become enthused about the scale and daring of a Moon mission, the level of teamwork involved, the idea of committing to a national project. After the first—suborbital—Mercury mission succeeded on May 5th, lobbing Alan Shepard up past the atmosphere and back down into the Atlantic without mishap, the die was pretty much cast. On May 25th, on live television, Kennedy told Congress and the nation that he believed that America should commit itself to landing a man on the Moon and returning him safely back to Earth within the decade. "No single space project in this period will be more impressive to mankind or more important for the long-range exploration of space; and none will be so difficult or expensive to accomplish."

He elaborated on the second part—on the fact that difficulty and expense were not a bug, but a feature—when he recommitted himself to the plan in a speech at Rice University, identifying Apollo with the "New Frontiers" he had promised in his 1960 election campaign. Putting the idea on a par with climbing mountains and Lindbergh's crossing the Atlantic, Kennedy told the crowd, "We choose to go to the Moon in this decade and do the other things, not because they are easy, but because they are hard; because that goal will serve to organize and measure the best of our energies and skills."

Initially, the preferred way to organise America's energies was a rocket even larger than those of the proposed Saturn family. The Nova would have a first stage powered by eight F-1s, a second stage powered by four F-1s. It would be large enough to put a fully fuelled rocket capable of coming back to Earth on the surface of the Moon in one go—a simple mission architecture, as such outlines are known, called direct ascent.

Even von Braun, though, came to think that the Nova was a bit much. Its rise might be so ferocious that it would have to be launched at sea from some sort of barge. The Huntsville team came to prefer an architecture that used two Saturn Vs—enormous rockets compared to everything that had gone before, but not as vast as the Nova—each launching part of the craft that would go to the Moon; the two parts would be mated in orbit. This was called Earth-orbit rendezvous.

It was an expediently stripped-down version of the way that von Braun had always thought such missions should be done. In popular articles in the 1950s he had described how the first step in human space exploration was to build a space station in low Earth orbit that could then be used as a base for assembling the craft needed to go to the Moon, or to Mars, and as a place to transfer crews between ships built to go up and down through the Earth's atmosphere and ships designed for the airlessness of space. Earth-orbit rendezvous was the same idea—but with the space station dropped so that all the assembly had to be done on the fly.

In the end, a third option won out: lunar-orbit rendezvous. Once the mission got to the Moon, it would leave the engines, propellant and heat shield needed for the return to Earth in lunar orbit. The astronauts would go down to the lunar surface in a little spacecraft designed specifically for that purpose. That reduced the amount of mass that had to go down to the Moon and, crucially, the amount that had to be brought back up.

This architecture approach allowed an entire Moon mission to be launched on a single Saturn V: a crewless service module with a largish engine, a conical three-man command module that sat on top of it and a lunar module (the LM, pronounced "lem") that would take two of the three-man crew down to the surface and back up. It was more complicated than direct ascent, because there were two manoeuvres needed in orbit. On the way to the Moon the command module, which needed to be at the very top of the rocket during takeoff if the crew was to have a chance of surviving a mishap, had to detach itself from the rocket's third stage, turn round, go back to the third stage and couple its nose to the top of the LM. When the LM came back up from the Moon, the same coupling had to be re-established so that the Moonwalkers could rejoin the pilot in the command module before the service module's engine sent all three of them home. Each manoeuvre was an opportunity for something to go wrong. But some slightly tricky piloting seemed a fair exchange for not having to build a monster like the Nova.

Without the lunar-orbit rendezvous, it is highly unlikely that Apollo would have made it to the Moon before the 1970s—or, quite possibly, at all. At the same time, looking back, it marked something of a loss. The minimisation of infrastructure that made it so appealing meant it left no legacy—there was no space station, as there might have been if a slower, more orderly approach to the Moon had been taken. Every Apollo mission would be a single shot. Once they were over, it would be in terms of hardware—even, to a degree, in terms of expertise—as if they had never happened.

No one worried about this at the time. They were doing something almost impossible—they weren't worried about setting up the sequel. Once they had shown what they could do, they would do more. Of course they would. Why wouldn't they? They would leapfrog again, on to Mars. They would build space stations after reaching the Moon instead of before—and cities in craters and new rockets powered by nuclear reactors and everything else the Space Age that was clearly dawning might need. Obviously, they would not just go to the Moon, look around, take note of the beauty of the Earth, pick up some rocks, come home and pack it all in. That would be madness.

● ○ ◐

APOLLO'S MEASURE OF THE BEST OF AMERICA'S SKILLS WAS IM-mense: by 1967 it employed some 400,000 people working through thousands of commercial and governmental entities. It was taking 4% of government spending (and this was while there was a war on). It was stretching the best minds in American aerospace to their limits and necessitating new ways of thinking and working across the continent—across the world, when you considered the telecommunications infrastructure required to keep track of the spacecraft.

But it was also intimate. Part of making lunar-orbit rendezvous work was making the spacecraft that actually went down to the Moon, the LM, as light as possible. In the original specification it was to weigh just ten tonnes. During development, it put on weight, despite furious attempts first to arrest and then to reverse the process. But it remained pretty tiny. And thanks to the need to carry fuel, oxidizer, life support, batteries, computers and more besides, the LM was noticeably smaller on the inside than the outside. The two astronauts had $4.7m^3$ of pressurised volume between them. That is roughly twice the volume of one of London's red telephone boxes.

Tiny. Also, a world. Or, at least, a fully functioning pinched-off little bleb of one. The LM gave the astronauts food and water; it kept their temperature stable; it protected them from meteorites. Its guidance computer mapped out their future. Once the LM was separated from the command module, it was all of Mother Earth they had left to them, save for voices on the radio: a microcosmic two-man planet.

A tiny world. But a fully functioning spacecraft, too—engines, guidance, communications, the lot. And one like none before it. Everything else on Apollo had, to some extent, been tried out at a smaller scale. There had been rockets fired by kerosene (in the first stage) and liquid hydrogen (in the second). There had been space capsules with heat shields for re-entry. But there had never been anything like the LM, something designed to come down from space and land under its own power rather than under a parachute. To land by its commander's hand and eye in a place where nothing had landed before.

And, although designed to land, also designed to be always in space. Previous spacecraft had had to carry their crews up through the buffeting atmosphere and bring them back down through it wreathed in fire. The LM's only duties with regard to atmosphere were to keep a very small one, composed of pure oxygen, contained within its tissue-thin aluminium walls (they flexed in and out as the air pressure inside them changed). The LM needed no streamlining, and when the first LM pilot, Rusty Schweickart, undocked the Apollo 9 LM, *Spider*, from the command module, *Gumdrop*, he was acutely aware of being in the first spacecraft ever to have been built with no heat shield. Dock again, or die.

The LM embodied a new off-kilter modernism—a form that followed function without compromise, however lopsided and implausible that made it look. The bottom half, to be fair, was fairly straightforward. It was a platform with an engine and legs—three in early designs, then five, then four. Octagonal, flat sided, its two fuel tanks and two oxidizer tanks arranged symmetrically around its central axis. Its job was to rob the LM of the velocity it would have when orbiting the Moon, allowing it to fall down to the surface, and to curtail that fall in such a way as to land at the designated site. Once on the Moon it was just a platform and a storage space with an all-important ladder running down one leg.

It was at the top of the ladder that function became complex and form became weird. The ascent stage had started off as a sphere, then been whittled down, then been added to. The result had a stubby-circular face like that of a somewhat satanic Thomas the Tank Engine: flattened nose, square eye sockets with deep-set triangular eyes, a round, shouty mouth. A fuel tank hung precariously off to the left like a goitre. Faceted like origami, aerials pointed in various directions, much of it was wrapped in gold foil to deal with thermal issues, obscuring its hard-to-follow lines yet further. There was just one concession to four-square order; at each corner, there were four rocket nozzles to steer with, one pointing up, one down, one forward or back, one to the side; x, y and z axes, as strictly Cartesian as the on-board computer required.

Inside, no seats. Room only for them to stand, side by side, looking out of the strange inset downward-canted windows, a throttle and joystick in front of each of them. A skylight over the commander—rank hath its privileges—and a small telescope, too. The hatch that led to the Moon knee-high between them, the inside of that angry mouth. No airlock. When they leave the LM, the whole thing is depressurized. Above the hatch, the DSKY—the guidance computer display and keyboard (numbers only—no qwerty). Above that, three more control panels. Spread around the rest of the walls, a dozen further control panels. One, in a rare stab at humour, is called ORDEAL: Orbital Rate Display, Earth and Lunar.

They stand in a well. At waist level the cabin opens out behind them in a raised alcove. At the top is the second hatch—the one that will let them back into the command module once they regain orbit. When they stand in the well, their helmets are in the alcove; when one of them needs to move around, he puts his helmet in the well. The personal life support systems that make their spacesuits self-contained— make the suits into leg-propelled spacecraft in their own right—are stowed to the side. So is the Environmental Control Atmosphere Revitalization Section which replenishes them, and which looks as if a madman had lashed drums of paint, plumbing valves, small fans and vessels for which there is no name into a framework of pipes and then applied a hydraulic vice to the whole assembly in every direction. Cram the flows and cycles necessary for life into the smallest possible volume and they have neither elegance nor any visual logic.

In the middle of the alcove is a squat cylinder like the continental-tire cubby on the back of a pre-war Oldsmobile, though not as wide across. It is the engine. In all the earlier spacecraft, the engine was somewhere else—strapped over the heat shield in the Mercury capsule, in its own separate chamber on the Geminis, the Vostoks and Soyuzes, the Apollo service module. In the LM it is right there in the middle of the crew space, tube-fed with fuel and oxidizer that are both toxic and explosive. There is a story that a LM fuel tank unwisely tapped with a ball-point pen during outdoor testing resulted in that pen being embedded in a fence post some way away, along with some of the unwise tapping finger.

During development, the fuel and oxidizer lines will not stop leaking. When Grumman ships the first purportedly flight-ready LM down to Cape Kennedy, it is rejected as not fit for the launch pad let alone for space: "Junk. Garbage." Trying to solve the problems makes the third LM so late to the Cape that there is not enough time to ready it for its scheduled flight.* What was expected to be a routine vacuum test for the fifth LM goes catastrophically wrong when one of the windows explodes.

The windows are crucial. There is a much-told tale that the first design for the Mercury capsules had no windows: the engineers saw no need for the astronauts to be able to see out, because they were basically just payload. Landing on the Moon, though, is not something that can be left to Ground Control—among other things, it takes radio waves just over a second to get there and just as long to get back. As Jack Myers, a life-support researcher at the University of Texas, put it at the time, "The human goes into space, not as a passenger, but as an essential part of the instrumentation needed for a particular mission."

The windows let the mission commander and the LM pilot, both of whom can land the craft, see what they are doing—they also connect them to the computer which turns the adjustments they make to joystick and throttle into digital instructions for the engines and thrusters. Born to give substance to science fiction's fascination with spaceflight in the context of a world reshaped by the arrival of science-fictional superweapons, Apollo added new depth to a third of the genre's concerns: new manifestations of intelligence and control in a world of thinking machines. The computer's requirements shaped the astronauts' world. For example: engraved on the inside and outside of the window glass is a sort of reticule. By holding his head so that the engravings on both sides of the glass line up with each other, the commander knows he is looking exactly where the computer thinks he is looking. That matters. The computer can respond to its human only if that "essential part of the instrumentation" is precisely aligned.

Computers on the ground also help with the windows' design. But this is the exception, not the rule; computer-aided-design software is not remotely up to handling the whole job as yet. All the

---

* That was why Apollo 8, originally intended to be a mission to Earth orbit using both the LM and the command module, became a command-module-only mission that went all the way to the Moon, and Rusty Schweickart and James McDivitt were the first to fly the LM on Apollo 9, an Earth-orbit shakedown cruise.

LM's complexities are drawn out by hand, and many are built by hand, too. The aluminium is so thin that it cannot be stamped into shape; it must be crafted. But computers are crucial, not just within the LM, but in the process of its creation. It organises. It measures. Software called PERT is used to schedule the development programme at Grumman, and most of the rest of the Apollo programme too, churning out new schedules every day, seeing what things that need to be done have not been done, what has to be done elsewhere so the next thing can be done here, marshalling an army of workers according to the planning procedures its programmers laid out for it.

Computers are the manifestation of the future that makes the future possible. They also make it visible, synthesizing experiences for which there is no prior experience. Flight simulators have been around since the early 1930s, when an enterprising young man called Edwin Link realised that the pneumatic systems his family used in their church-organ business could adjust the attitude of a pseudo-cockpit as if it were in flight. Having become widespread in the Second World War, this technology reaches its pinnacle in the Apollo simulators. Nothing has ever been simulated in advance remotely as thoroughly as the Apollo missions: the hours of simulator training run into thousands. In the LM simulators, computers coordinate instructions from the throttle and joystick with the movement of tiny fibre-optic cameras over plaster models of the lunar surface that would have made James Nasmyth deeply envious, thus showing the pilots the relevant bits of the Moon as they learn how to control their strange new craft under all conditions.

The need for such simulation pushes the computers into new virtual realms. The flight hardware needs to be re-created in ground-based software so that the simulators respond just as the real craft will. Virtual machines that exist only as lines of code run programs designed for real machines just as the real machines would—or so it is hoped. No one has made machines of pure logic before. As the programme goes on, some of the pilot's experience becomes purely virtual, too. The LEM Spaceflight Visual Simulator, created by General Electric in 1964, responds to the pilot's commands simply by moving pixels round a screen. In doing so it creates the first virtual landscape: no animated drawings, no plaster models, just zeroes and ones. At first it is purely geometric; with time it develops relief and shading. The technique starts to be used to explore different sorts of places, other sorts of travel. What would someday become cyberspace, and after that just the way that all images are created, starts off as a new way of showing the Moon to those about to walk on it. The prospect of an unprecedented physical experience brings forth a new virtual one.

Within these new directions of abstraction, though, intimacy remains—nowhere more than in the suit. Preconceptions suggested that the suit would be hard cased, with articulated arms—that it would make a man look like a robot. It is not. It is made of soft fabrics sewn together by women working with Singer sewing machines not unlike those found in half the houses of America, working not for a defence contractor but for the International Latex Corporation, makers of Playtex bras and girdles.

The spacesuit is the world shrunk skin-tight, the world three times removed. From the warm air of Florida to the command module; from the command module into the LM; from the LM into the suit. Sealed away airtight each time, and at the end of it all the breathable world is just in a bowl around the head and a pack on the back. The suits are better fitted to the wearers than any garment ever, sewn to an accuracy defined with aerospace exactitude, no stitch to be further than 1/64th of an inch—two-fifths of a millimetre—from the defined line of the seam. Not all the 21 layers are sewn; 16 of them, latex and Mylar, Dacron and Kapton, are glued together, no wrinkling allowed, the top layer almost indiscernibly larger than the bottom one, since what is outside must always be bigger than what is inside. Undergarments are webbed with waterfilled tubes to cool the skin; in the bright Sun with no flow of outside air to carry heat away there is always the risk of overheating. But warmth can be provided, too, as required. A different tube takes water to the mouth; another grips the cock to drain it away. That tube eventually comes in three sizes: large, extra-large and extra-extra-large; the first run, in small, medium and large, unaccountably saw some astronauts fitted with the wrong size.

As that shows, the suits, made by women, are for men. Astronauts were test pilots, and test pilots were men. Women could pass the same tests—and did, when they were applied privately and not by NASA—but they were not test pilots nor fighter pilots, and astronauts were. Some questioned this. Not many, though, and not high up. When Kennedy had said "a man on the Moon" it was not shorthand for a human of either gender. Such things were what men did.

As well as being men, the astronauts were white, too, white as the spacesuits.* That was not quite such a done deal. The White House knew that a black astronaut could be a big win, at home and abroad; it edged NASA in that direction, ensuring that there should be a black candidate in the next class of Air Force test pilots. The politicians did not, though, push the point when he was not selected for astronaut training. The first African American astronaut flew in only 1983, the same year as the first female American astronaut—who headed to space in the shuttle 20 years and two days after Valentina Tereshkova took off in *Vostok 6*.

Backing out of the angry-mouth hatch and down the ladder, the cycles of their lives wrapped around them, the men of the LM step onto the Moon. In a way they never reach it. Cocooned, drained and diapered, they are swaddled in the world they came from and return to.†
They do not feel the lunar temperature—they have their own. They do not breathe the Moon, or pee on it, or truly touch it; the gauntlets are

---

* As part of an art practice that interrogates notions of fantasy, modernity and that which it is to be African, Yinka Shonibare has made various spacesuit sculptures from the colourful patterned batik fabrics widely associated with West Africa. A black British security guard at London's Tate Modern who was spending a lot of time with those pieces once told my wife that he knew they were empty—but he felt a really strong urge to try to open their dark-glass visors and see if there was a face like his inside.

† In early planning, the idea of concentrating and controlling the astronauts' diets sufficiently strictly to avoid defecation during the voyage was considered. The borderline traumatic bowel movements suffered by test subjects at the end of ten-day trials of this approach, though, led to it being abandoned.

wonders of dexterity, given their thickness, but they cannot transmit the tactile. They can hear only themselves, and the voices of others, far away.

But for a few hours or days, depending on the mission, they inhabit it. They move back and forth across it, they jump above it and feel the light shock of landing in their knees as their muscles absorb their body's momentum.

They feel time pass on it. Though the Sun hardly moves in the sky, their hearts are beating, their reserves depleting.

They watch it respond to them; they see its surface pierced as they dig their trenches, and what they see matches what their muscles feel. They see its soft contours, pocked surface, hard-to-judge distances and near horizons in the way you see places that you may or may not go to while visiting nearby, not as you see things to possess, not as you see representations, or illusions, or other people's points of view.

It does not see them. And they do not see each other, at least not their faces. The sun-screening gold of the helmets' faceplates means no expressions make it out of the suit. Looking at each other, they see in the faceplates only pictures of the Moon, just as we do in the pictures they take of each other and bring back. They see what Moonwatchers have always seen: reflections. They see themselves.

They only experience the Moon in the flesh after regaining the LM. They bring its dust and grit in with them on their suits. They smell it in their air when the tiny volume of the LM repressurises and the helmets come off—it smells like gunpowder, or ashes doused with water. Sharp, electric sensations from reactions that could never take place in the vacuum outside catalysed in the air within.

The fine Moonstuff that coats the interior is dirt. It is pollution, in the way that the anthropologist Mary Douglas defined the word: matter out of place. Matter from the unworld in a new world.

In the LM, before he walks out into the dust, Buzz Aldrin takes Communion with bread and wine consecrated on another planet. "'I am the vine'," he says, "'You are the branches. Whosoever abides in me will bring forth much fruit. Apart from me you can do nothing.'" It is not the only lunar sacrament. In her book "The Planets" (2005), Dava Sobel recalls hearing that her friend Carolyn, on being presented with a speck of moondust by a planetary scientist boyfriend, impulsively ate it. The Apollo astronauts ingest it without choosing to. In their dust-dirtied LM tiny particles move through the alveoli of their lungs and across the microvilli of their guts into their blood, tissues and cells. They bring the Moon home incorporated. They bring themselves home changed.

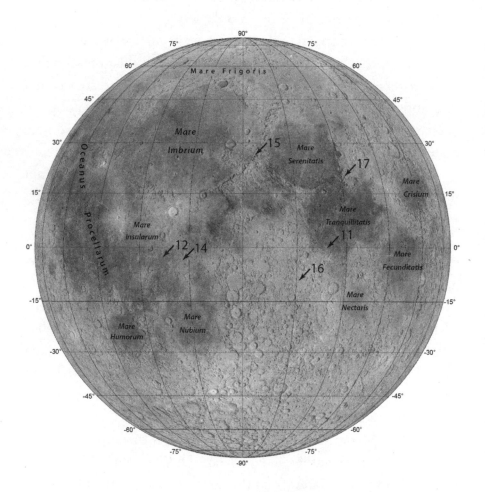

The Apollo Landing Sites:

11, Mare Tranquillitatis - Sea of Tranquility
12, Oceanus Procellarum - Ocean of Storms
14, Fra Mauro
15, Hadley Rille and Montes Apenninus
16, Descartes, Cayley Formation
17, Taurus - Littrow Valley

APOLLO 11: SEA OF TRANQUILITY

On the Moon: Neil Armstrong and Buzz Aldrin

In lunar orbit: Michael Collins

TIME SINCE LAUNCH 102:45:58

—Engine arm is off.

Houston, Tranquility Base here. The *Eagle* has landed.

—Roger, Twan . . . Tranquility. We copy you on the ground. You got a bunch of guys about to turn blue. We're breathing again. Thanks a lot.

—Okay. Let's get on with it.

103:03:55—THE VIEW

—The area out the left-hand window is a relatively level plain cratered with a fairly large number of craters of the 5- to 50-foot variety; and some ridges small, 20, 30 feet high, I would guess; and literally thousands of little, 1- and 2-foot craters around the area. We see some angular blocks out several hundred feet in front of us that are probably two feet in size and have angular edges. There is a hill in view, just about on the ground track ahead of us. Difficult to estimate, but might be a half a mile or a mile.

109:19:16—STEPPING OUT

—Okay. Houston, I'm on the porch.

—Roger, Neil.

109:23:38

—I'm at the foot of the ladder. The LM
footpads are only depressed in the surface
about 1 or 2 inches, although the surface
appears to be very, very fine grained,
as you get close to it. It's almost like a
powder. Ground mass is very fine.

Okay. I'm going to step off the LM now.

That's one small step for man;
one giant leap for mankind.

109:43:16—TWO MEN ON THE MOON

—Beautiful view!

—Isn't that something!
Magnificent sight out here.

—Magnificent desolation.

[silence]

Looks like the secondary strut had a little
thermal effects on it right here, Neil.

110:13:42—WALKING

—You do have to be rather careful to
keep track of where your center of mass
is. Sometimes, it takes about two or three
paces to make sure you've got your feet
underneath you.

About two to three or maybe four easy
paces can bring you to a fairly smooth
stop. [To] change directions, like a
football player, you just have to put a foot
out to the side and cut a little bit.

The so-called kangaroo hop does work,
but it seems as though your forward
mobility is not quite as good as it is in the
more conventional one foot after another.
As far as saying what a sustained pace
might be, I think that one that I'm using
now would get rather tiring after several
hundred feet. But this may be a function
of this suit, as well as the lack of
gravity forces.

110:16:03—CEREMONY

—All right. Go ahead, Mr. President.
This is Houston. Out.

—Hello, Neil and Buzz. I'm talking to
you by telephone from the Oval Room
at the White House, and this certainly
has to be the most historic telephone
call ever made. I just can't tell you how
proud we all are of what you have done.
For every American, this has to be the
proudest day of our lives. And for people
all over the world, I am sure they, too,
join with Americans in recognizing what
an immense feat this is. Because of what
you have done, the heavens have become
a part of man's world. And as you talk to
us from the Sea of Tranquility, it inspires
us to redouble our efforts to bring
peace and tranquility to Earth. For one
priceless moment in the whole history
of man, all the people on this Earth are
truly one; one in their pride in what you
have done, and one in our prayers that
you will return safely to Earth.

—Thank you, Mr. President. It's a great
honor and privilege for us to be here
representing not only the United States
but men of peace of all nations, and with
interests and the curiosity and with the
vision for the future. It's an honor for us
to be able to participate here today.

110:24:11—SOUND CHECK

—Buzz, this is Houston. You're cutting
out on the end of your transmissions. Can
you speak a little more closely into your
microphone? Over.

—Roger. I'll try that.

—Beautiful.

—Well, I had that one inside
my mouth that time.

—It sounded a little wet . . .

121:40:45—AFTER THE REST PERIOD

—Good morning, Houston.
Tranquility Base. Over.

—Roger. Loud and clear. And how is the
resting standing up there? Did you get a
chance to curl up on the engine can?

—Roger. Neil has rigged himself a really
good hammock with a waist tether, and
he's been lying on the ascent engine cover,
and I curled up on the floor.

124:21:54—LAUNCH

—9, 8, 7, 6, 5, Abort Stage, Engine Arm,
Ascent, Proceed.

We're off. Look at that stuff go all over the place. Look at that shadow. Beautiful.

26, 36 feet per second up.

—The *Eagle* has wings.

APOLLO 12: OCEAN OF STORMS

On the Moon: Pete Conrad and Alan Bean

In lunar orbit: Richard Gordon

TIME SINCE LAUNCH 115:22:16

—Whoopie! Man, that may have been a small one for Neil, but that's a long one for me . . .

I'm going to step off the pad . . .

Mark. Off the . . . Oooh, is that soft and queasy . . .

Hey, that's neat. I don't sink in too far. I'll try a little . . .

Boy, that Sun is bright. That's just like somebody shining a spotlight in your hand.

Well, I can walk pretty well, Al, but I've got to take it easy and watch what I'm doing.

Boy, you'll never believe it. Guess what I see sitting on the side of the crater!

—The old *Surveyor*, right?

—The old *Surveyor*. Yes, sir. [Laughs] Does that look neat! It can't be any further than 600 feet from here. How about that?

—Well planned, Pete.

115:27:27—NOT FALLING OVER

—Say, Houston; one of the first things
that I can see, by golly, is little glass beads.
I got a piece about a quarter of an inch
in sight, and I'm going to put it in the
contingency sample bag, if I can get it.
I got it.

Am I really leaning over, Al?

—You sure are. On Earth, you'd fall over,
I believe.

—Huh?

—On Earth, you'd fall over leaning that
far forward.

—It seems a little weird, I'll tell you.
Don't think you're going to steam around
here quite as fast as you thought you
were.

137:39:26—CONTENTMENT

—We're just sitting here now; we've got
the spacecraft all squared away. I'll say
everything's tied down, but man,
oh man, is it filthy in here; we must
have 20 pounds of dust, dirt, and all
kinds of junk.

—Roger, Pete. That'll be an interesting
zero g.

—Right. Al and I look just like a couple
of bituminous coal miners right at the
moment.

[Silence]

—But we're happy . . .

APOLLO 14: FRA MAURO

On the Moon: Al Shepard and Edgar Mitchell

In lunar orbit: Stuart Roosa

TIME SINCE LAUNCH 135:08:17

—Houston, while you're looking that up, you might recognize what I have in my hand is the handle for the contingency sample return; it just so happens to have a genuine six iron on the bottom of it. In my left hand, I have a little white pellet that's familiar to millions of Americans. I'll drop it down. Unfortunately, the suit is so stiff, I can't do this with two hands, but I'm going to try a little sand-trap shot here.

—You got more dirt than ball that time.

—Got more dirt than ball. Here we go again.

—That looked like a slice to me, Al.

—Here we go.

Straight as a die; one more.

[Silence]

—Miles and miles and miles.

APOLLO 15: HADLEY RILLE

On the Moon: David Scott, James Irwin

In lunar orbit: Alfred Worden

—Oh, look back there, Jim! Look at that. Oh, look at that! Isn't that something? We're up on a slope, Joe, and we're looking back down into the valley and . . .

—That's beautiful.

—That is spectacular!

Get the antenna pointed here.

—And it probably is fresh; probably . . .

—Okay.

— . . . not older than three and a half billion years.

—Can you imagine that, Joe? Here sits this rock, and it's been here since before creatures roamed the sea in our little Earth.

—Well said, Dave . . .

—Hey, Jim?

—Yeah.

— . . . well said.

—We ought to check the dust on the lens of these cameras.

145:41:48—THE GENESIS ROCK

—Okay, there's a big boulder over there down-Sun of us, that I'm sure you can see, Joe, which is gray. And it has some very outstanding gray clasts and white clasts, and oh, boy, it's a beaut! We're going to get ahold of that one in a minute.

—Okay, I have my pictures, Dave.

—Okay, let's see. What do you think the best way to sample it would be?

—I think probably . . . Could we break off a piece of the clod underneath it? Or I guess you could probably lift that top fragment right off.

—Yeah. Let me try . . .

Yeah. Sure can. And it's a . . . a white clast, and it's about. . . .

—Oh, man!

—Oh, boy!

—I got . . .

—Look at that.

—Look at the glint!

—Aaah.

—Almost see twinning in there!

—Guess what we just found.

—[laughter]

—Guess what we just found! I think we found what we came for.

—Crystalline rock, huh?

—Yes, sir. You better believe it.

—Yes, sir.

—Look at the plage in there.

—Yeah.

—Almost all plage.

—[garbled]

—As a matter of fact

[laughter]

Oh, boy! I think we might have ourselves
something close to anorthosite, 'cause it's
crystalline, and there's just a bunch . . .
It's just almost all plage. What a beaut.

APOLLO 16: DESCARTES

On the Moon: Charlie Duke and John Young

In lunar orbit: Ken Mattingly

TIME SINCE LAUNCH 124:55:39

—Hey, that LM makes a nice looking
house.

—Especially since it's about the only
one there.

—Yeah.

—You're right, Tony. It ain't nothing
much up here but a lot of rocks.

—Hope the door opens, Charlie.

124:56:58—RACING WITH THE ROVER

—Man, you are really bouncing!

(Pause)

—Is he on the ground at all . . . ?

—Okay; that's 10 kilometers.

Huh?

—He's got about two wheels on the
ground. There's a big rooster tail out of
all four wheels. And as he turns, he skids.
The back end breaks loose just like on
snow. Come on back, John.

And the DAC is running. Man, I'll tell you, Indy's never seen a driver like this.

Okay, when he hits the craters and starts bouncing is when he gets his rooster tail. He makes sharp turns. Hey, that was a good stop. Those wheels just locked.

APOLLO 17: TAURUS LITTROW

On the Moon: Gene Cernan and Harrison Schmitt

In lunar orbit: Ronald Evans

TIME SINCE LAUNCH 118:08:02

—Oh, man. Hey, Jack, just stop. You owe yourself 30 seconds to look up over the South Massif and look at the Earth.

—What? The Earth?

—Just look up there.

—Ah! You seen one Earth, you've seen them all.

145:26:25—PYROCLASTICS

—Wait a minute . . .

—What?

—Where are the reflections? I've been fooled once. There is orange soil!!

—Well, don't move it until I see it.

—It's all over!! Orange!!!

—Don't move it until I see it.

—I stirred it up with my feet.

—Hey, it is!! I can see it from here!

—It's orange!

—Wait a minute, let me put my visor up.
It's still orange!

—Sure it is! Crazy!

—Orange!

—I've got to dig a trench, Houston.

—Copy that. I guess we'd better
work fast.

—Hey, he's not going out of his wits.
It really is.

—Is it the same color as cheese?

170:41:00—DEPARTURE

—Bob, this is Gene, and I'm on the
surface; and, as I take man's last step from
the surface, back home for some time to
come—but we believe not too long into
the future—I'd like to just [say] what I
believe history will record. That America's
challenge of today has forged man's
destiny of tomorrow. And, as we leave the
Moon at Taurus Littrow, we leave as we
came and, God willing, as we shall return,
with peace and hope for all mankind.
"Godspeed the crew of Apollo 17."

ITS SURFACE

ITS SURFACE IS NOT UNIFORM. BUT ITS VARIATIONS ARE SUBTLE. On Earth, the surface is constantly re-created by cycles operating at every scale, from the deep-time dancing of tectonic plates to the freezing and thawing of dampness in the soil's pores. Molten magma rises imperiously, shouldering old strata aside; sediments shift and are eroded away. Sand blows into wandering dunes, flour-fine dust into fields of loess. There are limestone pavements, mudflat estuaries, abyssal plains. On the surface of the Moon, there are just dry rocks, jumbled.

The lunar regolith—from the Latin, "broken rock"—is just that: a blanket of rock fragments of every size which covers the whole surface. It is the product of billions of years of bombardment that has shattered and redistributed once-solid crust. Such impacts scatter dust. In the absence of air to resist their passage, motes of dust move as far and fast as boulders. Everywhere you find fragments of rocks from elsewhere—indeed, from everywhere, since the biggest impacts throw ejecta all around the Moon.

Only a few processes on Earth can carry a big rock a long way. The rocks thus moved are called erratics, and treated as telling rarities, clues to a past of glaciers, tsunamis and the like. On the Moon the erratic is the rule. The regolith is always a mixture of the nearby and the far flung.

The local, though, predominates. The regolith in the highlands is mostly made of anorthosite, the rock from which the Moon's primordial crust was formed, though it is laced with bits of basalt from the maria. The basically basalt regolith of the maria has a rather heavier admixture of anorthosite on the same basis. There is a smattering of other rock types, too. Not all molten rock that rises through the Moon's crust comes to the surface and ends up as basalt—some is trapped on the way, forming igneous intrusions of various types. Big impacts throw those rocks to the Moon's four quarters, too.

The Moon has no sedimentary rocks as such. Lacking flows of any fluid other than magma and lava, it cannot sort out particles into silts and sands and cobbles and lay them down in beds. Nor could it bury such beds and cook them into something new, as the restless Earth does. The nearest it comes to such transforming creativity is breccia, a rock made when the shock from a nearby impact fuses existing rocks and fragments of different sizes together.

Once formed, these breccias get slowly beaten down and broken up in their turn.

Beneath the regolith, the bedrock, too, is shattered—but into larger pieces. This fractured basement, which extends kilometres into the crust, is called the megaregolith. The boundary between it and the regolith proper is somewhat arbitrary and depends on the region's age—which is to say, the duration of its pummelling. In the ancient highlands the regolith may be 10 or 15 metres thick; in the younger maria it is more like five. On the very youngest surfaces, such as the floor of Tycho, only about 100m years old, the regolith may be only a few centimetres thick; below it lies a sheet of melted rock created by the impact itself.

The fact that impacts determine the nature and features of the surface does not mean that appreciable ones are common. If you staked out a square kilometre of the Moon's surface for close observation you would have to wait a few centuries to see it hit by something of a gram or more in mass. Impacts of inappreciable size, though, are constant. That square kilometre is hit by micrometeoroids a ten-thousandth of a trillionth of a gram in mass a hundred times a second or so. Each impactor would be only a couple of thousandths of a millimetre across—roughly the size of a bacterium. But again, in the absence of air, the little ones move just as fast as their bigger brethren, and their impacts have similar immediate effects, just on a far smaller scale.

Like the asteroid which formed Tycho, an incoming micrometeoroid digs out a crater a lot bigger than it is, melting some of the rock as it does so. This melted rock freezes back into a solid so quickly that instead of forming crystalline mineral grains, as lavas do, its constituents freeze higgledy–piggledy into glass. This glass sticks adjacent dust particles together.

There is thus a limit to the brokenness of the regolith, a steady state of creation and destruction.

- IV -

BOUNDARIES

Project Apollo was imagined as breaking a boundary in time as well as space. A world where people lived beyond the limits of Earth's atmosphere and gravity would be a world that had entered a new age: the Space Age. But the great push outwards achieved, instead, a turning inwards. The pocket turned inside out. The anti-Copernican revolution of "Earthrise" recentred the cosmos on closer concerns.

Like Odysseus back to Ithaca, those whose imaginations travelled to the Moon with Apollo returned not to some safe imagined home of the past but to one degraded, not to bright, confident Camelot, but to Watergate. America was losing a war and prey to inflation; its environment felt despoiled; its oil supply proved vulnerable.

Apollo was not universally admired. Both its ambitions and its costs brought forward critics as diverse as the philosopher Hannah Arendt, the artist Yves Klein and the sociologist Amitai Etzioni, the man who invented the term "Moondoggle". To Marvin Gaye, Apollo was of a piece with Vietnam, pollution, racist policing, the inner-city blues:

"Rockets, moonshots; give it to the have-nots". To most, though, including many who shared at least some of those reservations, it was still inspiring in prospect and exciting in its execution.

Afterwards, though, Apollo quickly became, at best, irrelevant. It had done nothing to clean the world, or feed the world, or take burdens from the shoulders of the world, or make the world more equal. Everything was the same or worse despite, or perhaps because, as Gil Scott-Heron angrily had it, "Whitey's on the moon". A bold, affirmative declaration—Yes, we can put a man on the Moon—became an inverted negative conditional, sometimes bemused, sometimes angry—If we can put a man on the Moon, why can't we . . . ? Cure cancer, or clean the air, or win a war in Indochina, or end poverty or curb inflation? Sure, these things are hard. But isn't hard what we do? Isn't hard what we *choose* to do? Isn't hard why we chose to do it? If we don't do these other things, is it because we choose *not* to? And who do you mean by we, white man?

The promise of Apollo was a high-water mark in terms of what a nation might aspire to do; the lack of any post-Apollo transformation, either in the heavens or on Earth, showed, for many, the limits of such aspiration. That is not to say the world did not subsequently change. But the change was not the deliberately brought about dawning of the Space Age. It was the unwilled unfolding of a new and troubling Earth Age.

Since Neil Armstrong stepped off the ladder at Tranquility Base, the Earth's population has increased by 100%, the amount of economic activity it supports by 300%; and the amount of carbon dioxide dumped in the atmosphere every year by 140%. About two-thirds of all the carbon dioxide emitted since the industrial revolution has been emitted in that past half century.

Over Antarctica, an ozone hole opened up. A fifth of the Amazon forest was lost. At the same time, fed by the extra carbon dioxide, the planet's plants put on a growth spurt. Seen from space, the Earth became perceptibly greener, and its red edge a bit sharper. Its seawater has become sourer, and the levels to which the Moon's tides can lift that water a little higher. Today's world is not yet as different from that of Apollo as Apollo's was from the deep past. But it is getting there.

An increasing number of scientists and concerned commentators refer to the new Earth Age as the Anthropocene—the age of humans. The conceit behind this naming, which dates from the turn of this century, is that the influence of humans on the Earth is no longer one factor among many but the single most important variable changing the way the planet works. As a result, the Earth, too, has crossed a boundary, entering a new period of geological time. In his influential paper "The Climate of History: Four Theses" (2009), the historian Dipresh Chakrabarty argues that the opening of the Anthropocene marks the point at which the Earth can no longer be treated just as a setting for human history, like the schoolroom globes that illustrators put in the Moon's sky before 1968. It has become the "Earthrise" planet, its dynamism increasingly shaped by human history and an increasingly active participant in it. Processes previously treated as parts of natural science—the carbon cycle, the rate of erosion, the evolution of the stratosphere—are now part of the political realm.

If this transition is to be treated as a formal transition in the Earth's chronology, as well as a tool of political and economic analysis, then the geologists who are custodians of that chronology need to find it in the rocks. Proposing the existence of systematic distinctions between older rocks from one age and younger rocks from another, and then disagreeing vehemently about where the line between them is best put, has been the bread and butter of geology for centuries. On this occasion the debate, formally taking place in the Anthropocene Working Group of the Subcommission on Quaternary Stratigraphy, a constituent body of the International Commission on Stratigraphy within the International Union of Geological Sciences, is being joined by voices from far beyond the neatly arrayed world of geology: philosophers, historians, environmental activists, among others.

There are four main options on offer. The original proponents of the idea of an Anthropocene liked 1750 or thereabouts: the beginning of the industrial revolution, and the baseline for all those graphs of rising temperatures and carbon dioxide levels. Others went back thousands of years further to the spread of methane-emitting rice paddies, the advent of farming or even the extensive use of fire. Some moved forward in time, rather than back, making a case for the period between the first atomic explosions, in 1945, and the last atmospheric nuclear tests, in 1963. Long-lasting radioactive isotopes laid down in sediments at that time provide just the sort of widespread, durable physical marker that geologists like when trying to distinguish befores from afters.

More recently, a small group has begun to push for the beginning of the 17th century, when carbon dioxide levels recorded in the polar ice caps wobble and pollen from New World corn appears in European lakebeds. Both markers have the same underlying cause: the Age of Being Explored by Europeans reached the Americas. Maize, along with many other foods, quickly spread across the Old World; measles, along with smallpox and influenza, spread across the New. These were changes big enough to be written in geology, too. Evidence of the new crops shows up in sediments across Europe and Asia. When some 50m people in the Americas died, new forests took root in what were once their fields, trees broke through the fallen roofs of their homes. The biosphere withdrew billions of tonnes of carbon from the atmosphere, investing it in trunk, root and leaf.

How do the versions of the Anthropocene defined by these dates differ in what they say about the relationship between the two histories that have become one? An Anthropocene which starts in the 1950s is a purportedly value-neutral just-the-facts-ma'am Anthropocene: its beginning marks the point at which, looking back, the strains on the Earth system first became apparent. It doesn't worry about why humans are having their impact, just notes that this is the well-marked point at which the scale of that impact started to increase very rapidly.

The steam-engine starting point says that what matters is the technology behind that impact: new ways of exploiting fossil fuels that brought with them the power to level mountains, create new chemicals, wage global war and support populations of previously impossible size. An earlier start makes things more natural again, if you are willing to extend your notion of nature to human nature. When apes get smart, control fire and learn to farm, they change the planet before you know it. The Anthropocene thus becomes a seemingly unavoidable consequence of the evolution of modern humans.

The 17th-century threshold says something more challenging: that the Anthropocene began not with a technology nor as a consequence of human nature but as embedded in history and politics—specifically in the appropriation of American natures and the dispossession of native American people, and in the creation of a global economy built on the accumulation of capital and an expectation of exponential growth.

The different dates all identify things that matter in different ways. And that is why I am moved to add to the mix a suggestion made by David Grinspoon, an astrobiologist, in his book "Earth in Human Hands" (2016): the base of the Anthropocene is to be found at Tranquility Base.

To take Armstrong's first footprint as the start of a new geological epoch would be to say very clearly that this epoch is of a very unusual kind. It is also to reinforce that the Anthropocene is a way of seeing as much as it is anything else—a way of seeing closely connected to the view of the Earth as a single and perturbable system that was given its iconic essence by the Apollo missions.

In making his case, Grinspoon points out that the *Eagle's* landing site handsomely satisfies the geologists' predilection for permanent markers distinguishing before and after: "The alien artifacts we left there will surely be detectible for as long as there is an Earth and a Moon." It is undeniably a sign of the human fixed in time and space. It also stems from the same conflict as the bomb-test sediments favoured as a marker by others and matching, Mr Grinspoon says, their "symbolic potency". Like them, it could only have been created by an entity that had "developed world-changing technology". As Verne suggested in "From the Earth to the Moon", the sort of technology that allows such travel is of its nature the sort of technology that is significant on a planetary scale.

Grinspoon's suggestion has the further benefit, at least to my eyes, that if Tranquility Base marks the bottom of the Anthropocene, then the Anthropocene is a geological epoch that encompasses both Earth and Moon. That seems at once odd and reasonable. If indelible human influence means the Earth has entered a new geological age, surely it means the Moon has, too. To be sure, what has been done to the Moon is beneath trivial compared to what has been done to the Earth. But the background rate of change on the Moon is so slow that the human novelties might still be seen to matter.

Apollo brought the Moon substances and processes it has never seen before. Never before has moondust been bathed in the exhaust gases of rocket landings and takeoffs—gases which, for brief periods, made up a substantial part of the Moon's ludicrously tenuous atmosphere. Never before has it had tire tracks traced across its surface or its boulders eroded by hammer blows. There is a sparse but real layer of human litter at six sites around the Moon, a strange sedimentation of abandoned experiments, blast-off detritus and sheer oddity—like the falcon's feather, dropped in tandem with a hammer, to illustrate Galileo's insight that, without air resistance, the two would fall at the same speed.

As yet these qualitatively unprecedented interventions do not quantitatively surpass even the Moon's very meagre natural processes of change, as some of the human influences that define the Anthropocene do on Earth. The mass of human relics and rubbish on the Moon is less than 10% of the 1,800 tonnes that hit it every year in the form of interplanetary dust. But 1,800 tonnes is less than the takeoff weight of four large airliners. The practical and political prospects for moonbases and colonies will be dealt with in later chapters (spoiler:

possible? Definitely, and on a smallish scale quite likely; large and/or enduring? Hard to say). But more than a few thousand tonnes moved on and off the Moon in a year is within the realms of feasibility. The traffic of supplies and personnel to and from America's McMurdo base in Antarctica is many times larger.

Though the geologists may look askance, an interplanetary Anthropocene also has the benefit of honouring one of the great 20th-century developments in their own science. Gene Shoemaker and his astrogeological colleagues showed that the stratigraphic reasoning of their science, the approach that underlay centuries of argument about the boundaries of eons and ages and epochs, applied beyond the Earth. The relative age of surfaces could be defined in terms of which rocks sat on top of which other rocks on the Moon as well as it could in Montana. Indeed, the impacts that made up the Moon's geological history lent themselves to stratigraphy from afar rather well. The ejecta from a large impact were often reasonably distinguishable, making clear a distinction between before—the surfaces they covered—and after—the ejecta layer and anything piled on top of it by subsequent impacts. Shoemaker's first geological map of part of the Moon established the impact that created the crater Copernicus as one such epoch-defining event. Today the history of the Moon is divided into five impact-punctuated periods in this way: the pre-Nectarian, the Nectarian, the Imbrian, the Eratosthenian and the Copernican.

The astrogeologists went on to apply a similar stratigraphic understanding to every planetary surface they saw, not to mention sundry moons and asteroids. Mars, Mercury and Venus all have geologic periodisations of their own. In doing this they also revealed the role that some of those distant objects have had in the history of the Earth— the history of cosmic battery which the Earth's face forgets but which the Moon's remembers. If geology applies elsewhere, why should some of the boundaries of geological time not do so, too?

One answer is that, before Apollo, what happened on one planet did not matter to the next. But this is not entirely true. There has been at least one other crucial event that links the geological history of the Earth to the Moon. And, as it happens, years before Grinspoon voiced his proposal for the Anthropocene, a quartet of scientists argued that that earlier event, too, should be recognised as a boundary in the geological histories of both bodies.

● ○ ●

AT THIS END OF THE GEOLOGICAL TIME SCALE, THINGS ARE WELL ordered. Until the Anthropocene is formally defined, if it ever is, humans live in the Holocene, a tiny sliver at the end of the 2.58m-year-old Quaternary Period, itself a subdivision of the 66m-year-long Cenozoic Era, the latest part of the Phanerozoic Eon.* The base of each is precisely defined—by a small but distinct climatic shift in the case of the Holocene, by the onset of the ice ages in the case of the Quaternary, by the thin level of iridium left behind by a dinosaur-killing asteroid in the case of the Cenozoic and, in the case of the Phanerozoic, by a 541m-year-old stratum in the cliffs of Fountain Head, on Newfoundland, just above which you can find the earliest fossilized burrows of a creature called *Treptichnus pedum.*†

At the other end of the time scale, unsurprisingly, things are considerably more rough and ready. The first of the Earth's four eons—a bookend similar in duration to the Phanerozoic but at the other end of the shelf—is known as the Hadean. It has no defined beginning; most people just sort of assume it began when the Earth did, around 4,540m years ago. It is widely held to give way to the subsequent Ar-

---

* Holocene means "wholly recent"; the Quaternary is stuck with the name it got as the fourth part of a chronological system no one uses anymore; Cenozoic means "the age of recent life"; and Phanerozoic means "the age of visible life"—ie, fossils big enough to be recognised without recourse to microscopy.

† That, at least, is where the International Commission on Stratigraphy puts it. Geologists in China and in Russia both argue for alternative sites and continue to use them as reference points. But all three are within a few million years of one another.

chaean 4,000m years ago, but that is basically just because people have got into the habit of saying so—there is no particular rock boundary anyone can point to and say, "This is the top of the Hadean, and that is the bottom of the Archaean, and here's why." Nor is it clear what event or change that boundary might mark.

In 2010, in a fit of playful tidy-mindedness, four scientists—Colin Goldblatt, Euan Nisbet, Norm Sleep and Kevin Zahnle—wrote a short paper trying to put some of this right. As it happens, I know and like all four of them—and, like many of their colleagues, I consider at least three of them to be both brilliant and a bit batty. Their 2010 paper, "The Eons of Chaos and Hades", reflects this. It is an attempt to stretch the geological time scale not just beyond the physical bounds of the Earth, as Grinspoon did, but back before its beginning. That it is undeniably fanciful. But it is not without sober sanction. No less an authority than the definitive "Geological Timescale" (1989) assembled by Brian Harland and others noted that "a pre-Hadean division to accommodate events prior to the earth's formation may be considered: but not in this work." And the events the proposed scheme seeks to fit into its framework are, as far as science can tell, things that must have happened in the sequence described, even if the actual dates for many of them are currently guesswork.

The story starts about 4.6bn years ago, at the point when a cloud of dust and gas which has begun to collapse in on itself due to its own gravity passes a threshold beyond which that collapse can but continue. It is the point at which the creation of a new star becomes a done deal. They make this moment of commitment the beginning of the Chaotian, an eon of swirling dust and gas.

They proceed to chop this first eon into two, the early Chaotian and the late. The boundary is "Let there be light". At the dense heart of the disk into which the cloud is still collapsing, the core of the about-to-be Sun becomes hot and dense enough for nuclear fusion to take place. Gravity produces pressures so high that smaller atoms are squeezed together into bigger ones; a chain reaction takes off in which the energy from one such fusion triggers the next and the next. Very quickly the Sun becomes bright—far brighter, in the exuberance of its birth, than it is today. The solar wind of charged particles that has streamed away from it ever since begins as a gale.

In the late Chaotian the Sun dims back down. The now illuminated matter swirling around it—which has, in the process of collapse, sorted itself out chemically so that the elements and isotopes present differ in different zones of the disk—clumps together into bigger and bigger lumps. Fairly soon some lumps are big enough and hot enough to undergo their own inner transformations; their centres melt, and iron, which does not like to be bound up in dust-made stone, sinks down to the core. These bodies now have stone mantles and iron hearts—they are, as the cosmochemists say, differentiated.

We see fragments of both the earliest undifferentiated bodies and the later differentiated ones on Earth: they fall from the sky today as meteorites. Through the cunning study of the various isotopes they contain their ages are known with remarkable precision. Some of the differentiated bodies become the planetesimals that Grove Karl Gilbert theorized about. These planetesimals went on to hit each other, too, forming what are now known as "planetary embryos". The bigger the objects, the more spectacular the collisions of accumulation.

In the outer solar system, where it is cool enough for volatile compounds such as water, methane and carbon monoxide to condense, these growing embryos wrap themselves in snow, ice and gas. The heavier they get, the more their increasing gravity can pull in; to those that have is more given. The biggest beneficiary of this positive feedback is Jupiter, which ends up weighing more than all the other planets of the solar system combined. Like a sun in miniature, it draws its own disk of gas and dust around it, a disk which produces four moons. A third of the age of the universe later, Galileo would trace their co-ordinated dance through his telescope, the first human eye ever to see them.

Towards the end of the Chaotian, the solar system is beginning to look quite familiar. Not all the planets have settled into quite the orbits they have today. But in the outer part Jupiter and the lesser gas balls are becoming recognisable, along with, farther out, billions of small icy bodies that never get swept up into anything bigger. In the inner part are almost-fully-assembled versions of Mercury, Venus and Mars, along with two other planets, Tellus and Theia. Theia is about the same size as Mars—which means about half the diameter of the Earth, and about a tenth of its mass. Tellus is about the size of Venus—almost as big as the Earth and about 90% as massive. It sits in an orbit very like that in which the Earth now sits.

And then, one day, something happens which, my friends say, brings the Chaotian to an end and gets the Hadean started. *One day*, here, is not a figure of speech; the boundary is as crisply defined as David Grinspoon's July 20th 1969 base of the Anthropocene. It, too, is a moment of contact, a meeting of worlds just like the touchdown of a spacecraft or a footprint on a grey-dust plain. A distinction defined by a coming together.

It is, though, on a larger scale. In one of the most violent acts in the history of the solar system, Theia collides with Tellus. The resultant mess eventually resolves itself into a new arrangement of mass and motion—a planet a bit bigger than Tellus, spinning rapidly, with a satellite much smaller than Theia circling it in an orbit only hours long.

The Chaotian is over. Tellus and Theia are gone. The Earth and the Moon have been born in their place.

● ○ ◉

A WEEK BEFORE THAT FATEFUL DAY, AS WE NOW RECKON TIME, Theia was the same size in the Tellurian sky as the Moon is in the Earth's sky today.* An hour before, it was as big as the dome of Saint

---

* Though the name Theia is widely used for the junior partner in the giant impact—Theia was the mother of Selene, the moon goddess, in Greek mythology—I should note that the name Tellus for the senior partner is not all that widely used. A lot of people just call it the Earth, or proto-Earth. But I think having a separate name helps, and Tellus is the one that the Chaotian quartet uses, so I have followed suit.

Paul's Cathedral seen from the north bank of the Thames, or that of the US Capitol seen from the pool that sits before it. The size of a Goodyear blimp, floating a few hundred metres above you.

Ten minutes before, it filled a third of the sky.

The land below may have been in day or in night. The incomer could have been crescent or gibbous, depending on the angle of the Sun. If crescent, swathes of its inverted night were bright-ashen-lit by the planet below. Mountain ranges hung down like damp-damaged paper from a ceiling, their shadows lengthening.

Or perhaps it came at midnight, totally eclipsed, an expanding absence of stars. Even then, it would have been something to see. The two planets' magnetic fields would have met hours before the impact. In a solar wind much stronger than today's, their mingling would have produced spectacular sheets and tangles of colour to light the land below and the land above.

Not that there was anyone there to see. Tellus and Theia may have had atmospheres and magnetic fields. Quite possibly they had oceans, too. They may even have had life. But if either of them did, it was simple, eyeless and probably deep below the surface. There were no birds to go silent in the sky, no animals to scurry and hide. No people to look up in terror as a ceiling spread over their world.

The airs and waters of Tellus would have felt the incomer's presence, though, through a gravity in the sky. Its waters would have risen up in unaccustomed tides: a metre or so a week out, but 20 times that a day out; in the final hours a mountain range of water reaches up towards Theia. The atmosphere, too, is stretching spacewards; a computer model might tell you from how fast, and from how far, the compensating winds raced in to replace the air pulled up—whether they created some sort of hurricane, as air rushing to an eye of low pressure does above hotspots in today's oceans. I cannot, myself, say. And minutes later, it would not have mattered. But I would like to know. Did the end which was a beginning come quietly or with the rushing of great winds?

Either way, it came. In the "standard model" of the Moon-forming impact—and what a story could be written about the catalogue of wonders that scientists litotically contrive to call "standard"—it came at 10km/s. That is fast: some ten times faster than a bullet, 30 times the speed of sound. But because planets are big, the collision itself is slow. At 10km/s, Theia merged with Tellus as slowly and surely as Armstrong would have sunk into the Sea of Tranquility had it been filled with Tommy Gold's quicksand dust.

More than ten minutes after the impact, there are parts of both planets which still do not know that it is going on. No shock wave, no blast of heat, no tidal wave or freakish jet stream has had time to reach them. It takes the shock waves almost 20 minutes to reach the mountain range of tidal water at the antipodes of Tellus; their encircling, narrowing, squeezing noose creates a waterfall jet up into space, like a child's cupped hands clapping in a bath.

It takes most of an hour for the planetary merger to reach its climax.

Slow does not mean gentle. The kinetic energy of a planet moving at 10km/s is vast—about the same amount of energy as the Sun puts out in a day or that the Earth receives from the Sun in 6m years. Before the impact had been going on for a second it had released more energy than all the world's nuclear weapons combined.

From that point of contact, the shock waves spread out as hemispheres, squeezing and heating the planets' mantles intolerably. Behind the shock came low pressure, which liquefied the trillions of tonnes of overheated rock in a flash. A conical sheet of red-hot magma hundreds, then thousands, of kilometres long shot out from around the point of contact. Within that cone, the leading hemisphere of Theia was quickly melting.

The planets' crusts were torn into fragments of every size, from mountain to Mozambique, some pushed out and aside, some crushed in the crunch of the mantles. As those disrupted mantles began to melt and flow, the planets' iron cores, themselves distorted by shock, found a new freedom. The cores were not headed straight for each other; Theia did not hit Tellus straight on, but at a tangent. Its core passed by that of Tellus, losing energy as it ploughed through the tortured mantle, the resistance of the rock—now partially molten, partially vaporised—stretching and streamlining and warping its iron from smooth sphere to mangled tear. It did not have the energy, though, to pass right through the rock and out the other side. It slowed and curled, turned and fell towards the core of Tellus. Within an hour the hammer hit the anvil, and sank into it. At the centre of the newborn Earth, the two cores became one.

The mantles did not quite coalesce as completely. As Theia struck its glancing blow part of its mantle sloughed off into that of the planet it had struck, but some drove on past, pushing a layer of Tellurian mantle ahead of it like mud on the blade of a bulldozer. Still travelling with more than half of Theia's original speed, blade and burden rose together back out into space. Much fell back. Much did not. Some escaped completely to form a short-lived ring around the Sun. But a lot stayed in orbit around the wrecked, reforming planet below. It was from that fiery orbital aftermath that the Moon was to grow.

The Earth writhes molten beneath its molten sky. Welcome to the Hadean.

●　○　●

WHAT COULD HAVE LED TO A SCENARIO SO EXTRAVAGANT BECOMing the most widely accepted account of the Moon's origin—albeit one which still has some big questions to answer? Its rise, which dates back to the 1970s and 1980s, was mostly due to the knowledge brought back from Apollo. Oxygen comes in three different isotopes. Apollo samples quickly showed that the ratios between these three isotopes in rocks from the Moon were very like those in rocks from the Earth—and unlike those of asteroids or of rocks from Mars (which sometimes fall to the Earth as meteorites, having been blasted off their home planet by much larger impacts). Such isotope ratios are taken to show in which zone of the great Chaotian disk the various rocks formed. Identical isotope ratios seemed to mean that the Earth and Moon formed in the same zone.

Analysis of the Apollo samples also revealed that moonrocks were very low in volatile compounds—water, carbon monoxide, nitrogen, sulphur and other light elements. Data from the seismographs the astronauts installed on the surface and from measurements of the Moon's gravity field made in orbit showed that it had only a very small iron core, if it has one at all. But if it formed where the Earth had formed—and thus presumably by the same mechanism and of the same stuff—how could that be? Why would it have so few of the volatiles with which the Earth was well stocked? Why had it not formed a big strapping core like the Earth's? Mars and Venus have done so. Mercury, the smallest but densest planet, has a core that takes up more than half its volume.

In short, in terms of its make-up, the Moon didn't really look like a planet in its own right. It looked like a dollop of the Earth's mantle which had somehow scooped itself out and placed itself into orbit, no core attached.

The idea of such a fission was first proposed by George Darwin, Charles Darwin's son. Darwin *fils* was interested in the drag on the Earth's rotation caused by the tidal bulges raised by the Moon. Turning beneath a tidal bulge that stays put, driving tides in and out of shallow seas and across great oceans, means that the Earth continually loses energy to friction—a loss which slows the Earth's rotation and lengthens the leash on which it holds the Moon.

This results from the conservation of angular momentum—a property of bodies, or systems, which depends on how their mass is distributed and how fast they are spinning. Move mass closer to something's spin axis and, if angular momentum is conserved, it will spin faster; move it farther out and things will slow down. It is an underappreciated side benefit of scientific progress that figure skating, originally developed on a frozen fen near Cambridge as a way of demonstrating this phenomenon, has gone on to become a very popular sport.[*]

You can change angular momentum only with a torque—a force applied off-centre so as to change the spin. With no torque applied to a system from outside, its angular momentum must stay the same.

This principle applies to the Earth-Moon system, tied together as it is by gravity. When the skater's arm extends, her body spins more slowly. The energy dissipated by the tides thus means both that the Earth's days are getting longer and that the Moon must have been closer in the past. By calculating the rate of its recession, Darwin found that, some 54m years ago, the two bodies would have been one. From this he derived the idea of a single body, spinning very fast, splitting into two. Long before it was explained by plate tectonics, devotees of Darwin's idea claimed that the great quasi-circular hole currently filled by the waters of the Pacific marked the divot from which the Moon had been thus ejected.[†] But no one could really explain why the planet would have come asunder in the first place.

The giant-impact theory, as the story of Tellus and Theia is known, seemed to provide an extraterrestrial precipitating event for the Moon's recession as well as explain everything else that other theories could not. It got a chunk of Earthish mantle, with its telltale oxygen-isotope ratios, into orbit without an iron core. It stretched out and melted that chunk, baking the volatiles out so as to ensure a desiccated end product. It even explained why the Earth-Moon system

---

[*] This is not true.

[†] Hugh Lofting made use of Darwin's fission idea in his book "Doctor Dolittle in the Moon" (1928), which may have been the first Moon narrative I ever read. As well as talking with the lunar animals—and plants—Dr Dolittle meets a caveman who has been living on the Moon since it was part of the Earth, rather as Cyrano de Bergerac meets the prophet Elijah when he visits the Moon's pre-lapsarian Eden. Lofting's sequel, "Doctor Dolittle and the Secret Lake", was probably my first exposure to the fiction of global cataclysm, though it is possible that honour should go to Tove Jansson's "Comet in Moominland".

had a high angular momentum in the first place. Theia's off-centre impact would have applied a massive torque to Tellus, producing a planet that spun very rapidly and the Moon that would, over billions of years of tidal braking, slow it back down.

Put forth after Apollo by, among others, Bill Hartmann—the man who first appreciated the ubiquity of ringed impact basins—and Don Davis, who helped guide Apollo 13 safely back to Earth, the giant-impact theory gained widespread credence in the mid-1980s. Early supercomputer models, some using code written to explore the effects of nuclear weapons, were able to sketch out what might have happened, an endeavour which seemed both sexy and confirmatory. But at the heart of the theory's success were the twin virtues of a great deal of explanatory power and no serious rivals. The idea of the Moon happening to pass by and being pulled into the Earth's orbit—the capture hypothesis—could not be made to work, then or now, without immense special pleading. Nor did it explain the similarities between the bodies. Co-accretion, in which the two simply formed together, explained the similarities, but not the differences—the Moon's lack of volatiles and core. Nor did it explain where all that angular momentum came from. The fission hypothesis lacked any sort of mechanism by which one planet might split in two.

What is more, the giant-impact theory helped explain one of the fundamental discoveries made by the Apollo missions. Whereas the dark plains of the maria were made of basalt, the brighter highlands were made of anorthosite, a rock composed mostly of calcium plagioclase, a mineral from the family called feldspars which are most familiar, I suspect, as the light-coloured non-quartz bits of granite kitchen worktops. If you take hot magma made from the Earth's mantle and let it cool under lowish pressures, calcium plagioclase is the first mineral to crystallise out as a solid.*

---

* The rock which Dave Scott and James Irwin, the Moonwalkers of Apollo 15, were so excited about on page 121 was a lump of near pure plagioclase; it has since been dubbed "the Genesis rock".

If formed from the orbital debris of a giant impact, the Moon would have started life covered by an ocean of magma—a hot layer of liquid rock hundreds of kilometres deep. (The post-impact Earth would have had such a magma ocean, too, but maybe only a tenth of the depth and possibly not over all of its surface.) As the ocean cooled, it did not freeze from the top down, as Nasmyth had argued it would. Minerals began to crystallise out at depth, the first of them plagioclases. Being lighter than the surrounding magma, they would have floated to the top. The magma ocean would thus have grown a crust composed mostly of calcium plagioclase.

Since the Moon, small and quick to cool, never developed any mechanism for recycling its crust, this primordial crust stayed put, except when blasted away by impacts or covered by later, darker basalts. One of the samples of almost-entirely-plagioclase highland rock brought back by the Apollo astronauts was 4.46bn years old—less than 100m years younger than the Earth and the Moon.

But for all its explanatory value—not to mention drama—the giant-impact theory has run into problems over the past decade. Further studies of moonrocks, using more and more delicate techniques to tease out finer and finer isotopic details, have found that they are not simply broadly similar to Earth rocks. In some respects, they are effectively identical. At the same time, more detailed computer models of the impact find that most of the stuff which would end up in orbit would have come from Theia, not Tellus. To reconcile this with identical oxygen-isotope ratios—and, now, some very detailed and similar measurements of the isotopes of other elements, too—would require Theia to be made of raw material remarkably similar to that of Tellus. If the two were identical to begin with, the explanatory edge gained by mixing them is blunted.

Few are ready to abandon the giant-impact theory because of this problem. That said, there is no widespread consensus on how to fix it. Some postulate a Theia very similar in composition to Tellus. Others prefer to make Theia either bigger or faster; that gets more energy into the system and gets more of the Tellurian mantle mixed in with Theia's and up into orbit.

In the early days of the giant-impact theory, higher energies were frowned upon, because unless you allowed a lot of special pleading they left the Earth-Moon system spinning too fast. Recently, though, mechanisms have been suggested whereby a torque applied by the Sun could bleed quite a lot of this excess angular momentum out of the Earth-Moon system quite quickly. The calculations on which this idea is based are not yet rock-solid; it is conceivable that, because it has the useful effect of allowing a wider range of impacts, the idea is getting a comparatively easy ride. For the moment, though, it has served to put high-energy impacts onto the agenda.

More energy means more mass in orbit, more heat, more angular momentum to stir things up with, more magma and a larger, hotter atmosphere of vaporised rock around the Earth. Indeed, the distinction between atmosphere and the mass in orbit could, to some extent, break down, creating an orbiting torus of molten and vaporised mantle much larger than the planet proper. Some of its proponents have started to refer to such a high-energy outcome as a "synestia", a thick doughnut-like disk dimpled in the middle. The Earth is in the dense middle of the dimple; the Moon will form out of the distended doughnut, which is a thoroughly stirred-up mix of both partners' mantles. Much of what does not end up in the Moon will return to Earth.

Whether you can hold such a cosmic doughnut together long enough to freeze a small planet out of it is an open question. The physics and chemistry that result when you pump a star-day of energy into a planet-sized rock are bound to be more complex than early models can grasp. But some way of getting more stuff into orbit and mixing it up more thoroughly seems, at the moment, a promising way to go.

And such strangeness is easier to think of now than it was in the early post-Apollo days, when a straightforward collision still seemed a touch outré. The discovery of thousands of planets beyond the solar system has stretched scientists' sense of what a planet can be. Some are so hot as to have their atmospheres permanently swollen, some locked so close to their star that one side is always almost melting. One star has a ring of hot rock around it that some have seen as the short-lived by-product of a collision quite as powerful as that of Theia and Tellus. The universe offers a far richer array of planetary possibilities than the bimodal distribution of small rocky inner and large gassy outer objects seen around the Sun today.

●　○　●

ACCEPTING, FOR THE TIME BEING, THAT THE MOON WAS BORN IN some sort of giant impact, was the fact that two planets hit each other in a way that would form a large moon unlikely? In some ways, such questions do not matter. It happened, or it didn't; look at the evidence, make the models, get new data and deal with it.* But from another point of view, it might be rather significant to know the answer, in a counter-Copernican sort of way.

The Earth has life—indeed, it has intelligent life. It also has a large moon. Is it possible that the two are related? If they are, then if a large moon is unlikely, planets with intelligent life may be, too. The Earth may be rare. These are the sorts of questions that keep astrobiologists up at night, often in bars.

In "Rare Earth" (2000), Donald Brownlee, an astronomer, and Peter Ward, a palaeontologist, make a vigorous and influential case that the Earth is unusual, and thus that humankind is too. Although microbial life might develop quite easily on many planets, they argue, the evolution of complex life had, on Earth, depended on various as-

---

* Measurements of oxygen-isotope ratios in rocks from Venus would be very handy; if they are Earth-like, then Theia could have been, too, and Mars is just an outlier. But getting rocks back from Venus is no easy task, and if there are meteorites from Venus on the Earth, they have yet to be identified.

pects of both its home planet and its home solar system being just so. The Moon is part of their argument.

The idea that the Moon had a relevance to life that goes beyond nocturnal illumination was not new. Some had argued that, without the Moon, the Earth would have a stiflingly thick atmosphere like that of Venus. Others had suggested that lunar tides—much bigger in the Earth's early days because the Moon was still much closer—were crucial to life's origins. By sloshing seawater into tidal pools from which it would then evaporate, they provided a way to concentrate the chemicals life would need. This is not an idea many people are interested in at the moment—recent fashion has been to look for life's origin in deep-ocean hydrothermal vents where the tug of the tides does not matter. But ideas on the subject have changed before, and may change again; hypotheses, like tidal pools, come and go.

Brownlee and Ward, though, plump for another lunar effect—a damping down of the Earth's wobbling. Planets do not sit up straight in their orbits: they lean over. The Earth's axis of rotation is currently at an angle of 23.4° to the vertical, as measured with respect to the plane of the ecliptic. It is slowly in the process of sitting up straighter; but once it reaches about 22.1°, in a bit more than 10,000 years, it will start to lean back over again. Its obliquity nods between 22.1° and 24.5° every 41,000 years. The effect this nodding has on the intensity of the planet's seasons is one of the things which sets the rhythm of the ice ages that mark out the Quaternary.

On more-or-less-moonless Mars, shifts in obliquity are both bigger and less regular. Sometimes Mars sits bolt upright, with no seasons worth mentioning. At other times it reclines as far as 60°—a posture in which its inhabitants, if there were any, would experience extraordinarily hot and cold hyper-seasons, with the midnight Sun seen far into the tropics at midsummer.

In the 1990s Jacques Laskar, one of the astronomers who discovered the role that chaos plays in the seemingly stable solar system, showed that the difference between the Earth's gentle nodding and

Mars's wild oscillations could be accounted for by the Moon. A constant lunar tug on the Earth's equatorial bulge—a paunchy distortion of the planet's sphericity caused by its spin—keeps it sitting up pretty straight. Take the Moon away, and the Earth's obliquity becomes even less stable than Mars's, swinging as high as 85°—a planet flat on its back. Having the poles point almost straight at the Sun during summers and almost directly away from it in winters would remove all temperate zones from the planet.

In "Rare Earth", Brownlee and Ward argue that these sometimes extraordinary obliquities would give a moonless Earth a climate so catastrophe-prone that complex life would be very hard put to flourish. Subsequently, though, the story has been shown to be a bit more complicated than that.

Changes in a planet's obliquity depend on the gravitational influences of the other planets in the solar system. The slower that planet rotates, up to a point, the more sensitive it is to these chaos-inducing tugs. The Earth and Mars are in somewhat similar orbits and currently have days of very similar length. That is why Laskar found that the axis of a moonless Earth jerks back and forth.

But unlike moonless Mars, which may have had pretty much the same length of day for all its history, the Earth has not. The Moon may be stabilizing the Earth's obliquity now—but as George Darwin pointed out, it is also responsible for the Earth having a sufficiently long day for chaotic obliquity changes to be a risk in the first place. If the Earth had started with a ten-hour day and no Moon, it would still have a ten-hour day, more or less, and its obliquity would have been stable all the while.

It is still possible to make a case that complex life is a lot more likely on an Earth-like planet if it has a big moon. David Waltham, a British astrobiologist, suggests in his book "Lucky Planet" (2016) that complex life needs both a pretty stable obliquity and a fairly long day—a combination the Earth would not have were it not for the Moon. On planets with significantly shorter days, he argues, the transfer of heat from equator to poles would be less efficient. The winds and currents responsible for that transfer are diverted from the direct equator-to-pole trajectory that you might expect into the looping swirls of brilliant white seen in "Earthrise" by the Coriolis effect, which swings them to the east or west. The faster a planet spins, the stronger that effect will be—and the harder it will be to get warmth to the poles. Dr Waltham argues that the Moon is just the right size to allow the Earth both a stable obliquity and poles warm enough to keep most ice ages relatively minor. It is a cunning argument, but not a compelling one. It may be that making significant progress on the question of the Moon's importance for life will have to wait until the presence of complex life is—or is not—discovered by inspecting the earthlight-like light from distant planetary systems.

If complex life can indeed develop on a world without a large moon, a further question arises: what would it have been for humans to play out their history under a unMooned sky? Something, surely, would have been lost: but what? Moonlessness is not in itself an untoward experience; it is, after all, a monthly phenomenon. New Moon and no Moon seem hardly different. But the dynamics of the world would change. Night would be a deeper thing, always dark, unchanging; the sea gentler, its tides low and metronomic, never spring and never neap. Nothing would wax or wane; the kneeling-god-drama of eclipse would be unknown: the seasons would be the only cycles, the constellations, the only permanent features of the sky.

Which is all to say that you would miss it, were it gone. But not that you would wish for it if it had never been. What madness it would be, to imagine the waxing, waning, night-lighting Moon in an unMooned world—to dream of a sky-thing which could slide seamless over the Sun. And that poses a further question. What might this world lack that is as hard to imagine as the Moon would be on an unMooned Earth? Of what absences are we unaware?

As well as an absence, the Moonless world would be marked by difference. Folklore, nocturnal action and assignation, marine and maritime life: all would be otherwise. Also, perhaps, science. In an essay published in 1972 Isaac Asimov argued that, for all that its craters and earthshine played a role in the acceptance of Copernicanism in the 17th century, in the greater scheme of things the existence of the Moon had thwarted the progress of astronomy, slowing the necessary move from a picture of the universe centred on the Earth to one centred on the Sun, for the simple reason that, although the other planets and the Sun do not orbit the Earth, the Moon really does.

The Sun's apparent path through the sky can be explained equally well by the Sun going round the Earth and the Earth going round the Sun. The same is true for the Moon, though, as far as I know, no one has ever actually put forward a lunocentric theory of the Earth's movement. But there is no way to explain the Moon's path through the sky in terms of the Moon orbiting the Sun. So either the Moon moves round the Earth and the Earth moves round the Sun, and the universe has more than one axis for things to turn on, or the Sun and Moon both move around the Earth. The second is obviously much simpler, and also satisfies the strong intuition that the Earth is stationary. A geocentric solar system made explaining the paths of the other planets through the sky a little tricky, but there were ways around that, if you were imaginative enough.

If there had been no Moon, Asimov argued, astronomers from Babylon onwards would have realized that having everything move around the Sun would be just as simple as having everything move round the Earth—and would have made understanding all the other planetary orbits entirely straightforward. If this truth had been known throughout history, he went on, there would have been much less conflict between science and religion—indeed the former might have wholeheartedly supported the latter, something he thought devoutly to be wished. The mechanistic, gravitational revolution associated with the period from Copernicus to Newton might have come about centuries or millennia earlier.

What was more, he speculated, a non-geocentric theory of the universe would have encouraged a less anthropocentric attitude to the living world, and thus no environmental crisis of the sort which, by that point, was a deep concern to him. Without the Moon, in short,

20th-century science might have been millennia in advance of where it was, and a Galactic Empire well on its way. With it, the world of the 1970s was close to a total ecological collapse. He called the essay "The Tragedy of the Moon".

The gloomy ideas are, in themselves, entertaining trifles. A historiography in which science simply springs up if not suppressed, and a philosophy in which astronomy has enough access to the human heart to govern its reverence for the world, hardly seem compelling. But that such thoughts came to one of America's greatest science fiction writers when, at the time of Apollo, he looked out across the early morning New York skyline at the setting Moon speaks of a pessimism worth noting. And there is also the question Asimov does not examine. The Copernican revolution might have come about more quickly if deprived of the Moon's empty landscape in the sky, but would it have been as profound? If the Earth had been treated as just another planet since the birth of astronomy, then how, and why, would the other Sun-circling points of light have become worlds?

●　○　●

THE CIRCUMSTANCES OF THE IMPACT THAT FORMED THE MOON may or may not have been unlikely, which may or may not have implications for life elsewhere. A catastrophic rain of later impacts, though, must most definitely have happened. It can be read on the Moon's face. But how long, and hard, was that rain?

The astrogeological effort that Shoemaker got under way in the 1960s created a relative time scale for lunar landscapes: the Copernican sat on top of the Eratosthenian, on top of the Imbrian, and so on. Providing dates for the transitions, though, was hard. The best tool to hand was, again, provided by impacts. Younger surfaces will in general have fewer craters on them than older ones. If you had a model of the rate at which things hit the Moon today, and estimates of how that rate might have changed over time, you could translate the frequency of craters into estimates of length of time the surface studded by those craters had been exposed to impacts. On this basis Bill Hartmann calculated in the mid-1960s that the lunar maria were about 3.6bn years old. When maria basalts brought back by the Apollo missions were precisely dated in the lab, they showed a range of ages pleasingly close to Hartmann's estimate. Unsurprisingly, the rocks associated with the impacts which had created the basins in which those basalts sat were older.

What came as a shock was that those impact-basin ages seemed to be highly concentrated: the impacts that had formed the basins seemed to have happened rat-a-tat-tat almost half a billion years after the Earth and Moon had formed. This led to the notion of a "Late Heavy Bombardment"—that some 500m years after the formation of the solar system the rate of impacts, which was in long-term decline, suddenly peaked back up for some reason. As astrobiology got going in the 1990s, this phenomenon began to look like a very interesting connection between the history of life and the history of the solar system.

If the bombardment was responsible for most of the visible damage to the surface of the Moon, it would have treated the Earth even worse; the Earth's greater size and stronger gravity mean that a given population of incoming rocks will hit it more often and harder. If there were 30 to 40 basin-forming impacts on the Moon, there would have been 100 or more on the Earth. The biggest of them would have been bigger than any on the Moon, capable not just of burning the land but also of boiling the sea: the water from all the oceans would have been turned into a thick atmosphere of superheated steam, and the crust sterilised down to a depth of a kilometre or more.

The earliest universally accepted evidence of life on Earth is 3.5bn years old. But there are rocks that date back 3.8bn years that are taken to carry strong chemical hints of the presence of life. If a whole spate of potentially planet-sterilising impacts took place 3.9bn years ago, that 3.8bn-year-old evidence suggests either that life got started very quickly or that life has a remarkable resilience. Both possibilities are interesting to astrobiologists.

If you think it took the whole early history of the Earth—in effect, all 500-odd-million years of the Hadean—for life to get going, then you will tend to think it a more unlikely process than if you think it got going in just 100m years; things that take a long time to come right seem intrinsically less likely than things which come off quickly. If life could get going in just 100m years, people argued, life might be quite an easy trick for a planet to pull off.

Perhaps more intriguing was the idea of resilience. In 1998 Norm Sleep, a professor at Stanford, and Kevin Zahnle, a researcher at NASA Ames—half of the Chaotian quartet—published a paper noting that if the Earth was subjected to ocean-boiling impacts, then one of the best ways for life to survive them was to follow the advice that protagonists in gangster films always ignore and get out of town till things cool down. Big impacts throw smaller rocks nearby out into space. A few travel to other planets, as the presence of meteorites from Mars on the Earth demonstrates. Most, though, eventually fall back whence they came. But they may spend hundreds, or thousands, or hundreds of thousands, of years in space before they do so—long enough for even the effects of an all-out land-burning ocean-boiler of an impact to have worn off.

Satellites brought back to the Earth from orbit have demonstrated that the inert spores formed by some bacteria can survive for years in space. Lodged into the pores of a rock they might survive for millennia. And if the orbital-refuge idea is true, this might be related to the fact that their most distant ancestors evolved to do so. Boiling the oceans represents quite the evolutionary bottleneck. If everything that can't survive a few thousand years in space is wiped out, then what repopulates the Earth will be space-ready by definition. Tsiolkovsky's notion of life evolving towards space would be given a twist; to leave the planet would be to revisit its childhood survival strategy.

Or even, perhaps, its birthplace. If primitive life can survive occasional episodes of spaceflight, then the origin of life need not be on the planet where that life takes root. The arguments that apply to terrestrial microbes surviving superheated atmospheres by sitting them out in space apply to Martian microbes, too, should there have been any. If the conditions of the early solar system encouraged life to develop the capability to survive spaceflight, in doing so they also provided it with the ability to pass from planet to planet.

In the early solar system, with lots of big impacts going on, the rate at which meteorites from one planet arrived at another was surprisingly high. The orbital refuge paper calculates that, back then, thousands of meteorites from Mars rained down on the Earth every year; some of those rocks would have been in transit for less than a decade. The flow of rocks from Earth to Mars would have been considerably smaller—it is harder to launch a meteorite off the heavier Earth, especially if you need to get it into space at a speed that will take it all the way to Mars. But it would still have been appreciable.

If life originated on Earth, it could have spread to Mars through this sort of "transpermia"; if on Mars, it might quite likely have fallen to Earth.*

Even while becoming part of the increasingly cosmic context in which people thought about early life, though, the Late Heavy Bombardment was not universally accepted: Hartmann, among others, never liked it. He thought that to the extent the effect was real, the lack of evidence for basins more than 4bn years ago simply showed that such evidence got written over, not that such impacts never happened. There was not an uptick in impacts around 3.9bn years ago; it was just that later impacts had overwritten the evidence of the earlier ones. This is an increasingly widely held view. Arguments that the bombardment might have been caused by Jupiter and Saturn swinging in towards the Sun and out again thanks to a peculiar orbital resonance, which seemed to provide a mechanism whereby the solar system would have

---

* The process is also known as "endolithic panspermia"—panspermia being the idea, dating back a century or so, that life spread across the galaxy in the form of spores, and endolithic meaning that in this case it did so inside rocks. I think my term, transpermia, is better in that it brings out the idea that this is life moving specifically from one place to another rather than being broadcast to everywhere.

been filled with asteroids and comets thrown out of earlier, stable orbits around the relevant time, look less convincing now than when
they were first made a decade or so ago.

As well as finding itself without an explanation, these days the
rock record of the Late Heavy Bombardment looks a little more dubi-
ous, too. Recent studies suggest that the Apollo rocks seemed to offer
evidence for impacts clustered around the same time simply because
most, maybe all, of the rocks in question actually came from the same
impact: Imbrium. Its debris is estimated to have covered about a fifth
of the Moon's nearside. Its remnants can be found in rocks from almost every Apollo landing site. Rocks taken to have come from other
impacts may have simply been Imbrium ejecta mischaracterized.

About ten years ago, when America's National Academy of Sciences put together a report on what science needed to be done on
the Moon, sorting out the timing and severity of the Late Heavy
Bombardment, if any, was top of the list: "Science goal 1a". That report concluded, as has almost everyone else taking an interest, that
this means getting some more Moonrocks, looking at the isotopes
that date them and thus establishing in absolute, rather than relative,
terms when various impacts happened. And the place to start is with
the biggest impact of them all (bar the one at the beginning), the one
which created the South Pole-Aitken basin. As the name suggests,
it stretches up all the way from the South Pole to the farside crater
Aitken—just 17° south of the equator. That makes it 2,500km across: a
whopper by any planet's standards. You could fit India and Argentina
into it and have room left over. If you were willing to leave off the
autonomous regions of Guangxi, Inner Mongolia, Tibet and Xinjiang
you could get all the rest of China into it.

Unlike the biggest nearside basins, South Pole-Aitken has no
smooth sea of maria basalt at its base. But it is still distinctly darker
than the surrounding highlands, probably because it has dug deeper
into the crust than any other basin. It is 13km deep; the Leibniz
Mountains, which define its north-eastern rim, are from foot to
peak the highest on the Moon. Stratigraphers have decided it is

the oldest distinctly identifiable feature the Moon has to offer. They
have also identified places in it where there may be rocks which were
melted in the impact itself and which would provide a precise date
for it.

Whether or not impacts ticked up just before and around the creation of the Imbrium basin, they may have dropped off quite quickly
afterwards. And so, in the spirit of continuing an interplanetary trend
in geological time scales, I suggest the following. Given that, as yet,
there is no specific marker for the end of the Hadean on Earth, might
it not, at least as a temporary measure, be fair to use Imbrium to
date the end of the Hadean too? Whether or not there was a spasm
towards the end, basin formation really was the dominant geological
process on the Moon and a very important one on the early Earth.
In the 3.8bn years afterwards they really did diverge—until humans
began to mess around with both of them in the very recent past.
Whether or not one agrees that Earth and Moon should share an
Anthropocene, it seems reasonable that the end of their distinct but
twinned childhoods should be marked by a single event—and if it is
to be one of which a clear record remains, that has to be an event on
the Moon. The event responsible for the most distinctively rimmed of
the Moon's seas, the arc of its half-encircling mountains visible to all,
is surely as good a one as any.

● ○ ●

IMPACTS HAVE CONTINUED FOR THE REST OF LUNAR HISTORY,
the only real exception to Robert Heinlein's tongue-in-cheek dictum
that "nothing ever happens on the moon". And they brought the Moon
at least two things that humans might treasure.

The first is water. Many asteroids are made of minerals that contain a bit of water; those known as "carbonaceous chondrites" can be
over 20% water by mass. Comets are a good bit wetter still. When
a body of either sort hits the Moon, the water it contains is vapo-

rised and much of that vapour is immediately lost back to space. But some sticks around. On the hot, sunlit side of the Moon's night-edge, it forms a tenuous atmosphere; on the dark, cold side, an all but undetectable frost. As the night-edge sweeps round the planet, the volatiles move from ice below to vapour above and back again on a monthly basis.

In time, most of this asymmetric atmosphere is lost—the Moon is too small to keep such a wrapping around it. The Sun's ultraviolet light ionises the volatile molecules, after which the charged particles of the solar wind strip them away. But some of them remain as frost in perpetuity—because some of the Moon never sees the light of day.

The Moon has a low obliquity; it sits almost straight up with re-spect to the ecliptic. This means that the Moon's poles are lit tangen-tially, with the Sun never rising far above the horizon. The shadows are long—so long that some of them never end. In craters at the poles there are places where the horizon-hugging Sun cannot shine. It may rise high enough to light the inner rim of a crater, creating the morning-lit side which Galileo, when first convincing people that the craters were craters, compared to the western side of an Alpine valley. And as the Moon slowly turns, the part of the inner rim that is lit changes, too, as if being broiled on a sluggish rotisserie. But though most of the rim is illuminated at some time or other, the floor never is. The only light it sees is the secondary light reflected from the rim.

And some of the crater's interior does not even see that—because there are craters within craters, and from those inner craters the rim of the outer one often is invisible. The depths of such craters see the Sun neither directly nor indirectly.

Most of the craters which contain this perpetual darkness are around the South Pole: the crater named after Gene Shoemaker is one. Being in the depths of the South Pole-Aitken basin gives the region a head start when it comes to avoiding sunlight. But there are pools of perpetual darkness in the north, too. And at both poles the darkness is phenomenally cold—colder, remarkably, than the surface of Pluto, which is 30 times farther from the Sun. Pluto may get sunlight a thousand times weaker than that which bathes the Moon, but every square metre of it gets some of that light some of the time. Go without sunlight at all for a few billion years and you can get really cold: the floors of the sunless craters are at about minus 238°C, 35 degrees above absolute zero.

If vapours produced by impacts or possibly from other sources rime these craters with frost and nothing subsequently re-vaporises it, it is fair to imagine that that frost accumulates. Such accumulation would make the creeping growth of glaciers look whip-tip fast; but it has had billions of years in which to play out. And so something very slightly like a glacier growing into the sky could take form: a laminate of dust-dirty ice, growing a few millimetres every million years if it's lucky, lit only by the stars towards which it is so very slowly reaching.

At least, that was the case for the past few billion years. Around the 25th year of Grinspoon's Anthropocene, though, other radiations began to impinge on the sunless craters. First radar, then lasers, shone down from orbit to probe their depths. Other instruments, rather brilliantly, made use of the stars themselves, picking up reflections of ultraviolet starlight from the craters' interiors. Together these and later studies provide strong evidence that layers of ice really do exist in the craters' depths.

This has made those enthusiastic about the Return to the Moon very happy. Layers of ice at the poles could be used to provide a research base, or indeed a permanent settlement, with water and oxygen, greatly cutting down on the need to bring supplies up from Earth. And splitting water into hydrogen and oxygen gives you high-quality rocket fuel and the perfect stuff to burn it with.

The other treasure that impacts bring to the Moon is rocks from elsewhere. The impacts that deliver meteorites from the Earth to Mars deliver far more of them to the surface of the Moon; and there will be meteorites from Mars, too, and from Venus. Random bits of all the inner planets are scattered over the lunar plains—mostly buried a bit below the surface, perhaps, thanks to the slow but ceaseless churn of new impacts. But some will almost certainly still be identifiable, with scrutiny.

In a beautifully titled paper from 2003, three planetary scientists suggested that this made the Moon "Earth's Attic": no one knows quite what is stored up there, and a lot of it is probably junk, but there is more of it than you might think, and it is older, too. There might be valuable oddities. There might be precious heirlooms.

On the valuable oddity front, there might be 30kg of rock from Venus on every 100km^2 of the lunar surface. Finding them would be a heroic task. But then, getting samples of Venus is always going to be a heroic task, what with its surface being at a temperature of 440°C and smothered in an atmosphere 100 times thicker than the Earth's. It is hard enough to land there; only two Soviet probes have done so, and they lasted no more than a few hours. Landing, picking up some rocks and getting back into orbit—which is almost as hard to do from the surface of Venus as it is from the surface of the Earth—is a challenge beyond today's technology. And even if it were possible, it would only bring back bits of today's fairly young, lava-covered crust. The crust from billions of years ago—back before the advent of that crushing atmosphere, back when Venus might even have been an ocean—would be unreachable.* But there may be some of it on the Moon, because most of the transfer of rocks between the inner planets took place in the heavy-hitting Hadean.

However much of Venus has ended up on the Moon, there will be a lot more of the Earth there. On the same 100km^2 patch where scientists might, if lucky, find 30kg of rock from Venus, the Earth's Attic theorists would expect 20 tonnes of earthrock. Again, most of it would be from the Hadean—a period the record of which on Earth itself is all but non-existent, thanks to the planet's constant recycling

---

* This possibility of a clement early Venus is, as it happens, a subject particularly associated with David Grinspoon.

of its rocks. In early 2019, geologists in Houston announced that they had identified what they took to be one such fragment in a breccia brought back by Apollo 14.

It is not just the practices of Earthly geology that have been exported into the solar system through planetary science, nor its subdivisions of time, be they Hadean or Anthropocene. Its oldest, dearest and rarest subject matter sits in the sky above, not the ground below. James Nasmyth was wrong about the Moon preserving the look of the early volcanic Earth, but he was right in seeing it as home to the vestiges of creation.

If there are findable rocks anywhere in the universe which contain traces of the earliest life on Earth, the odds are that they are to be found on the Moon.

A UNIFIED GEOLOGIC TIME SCALE

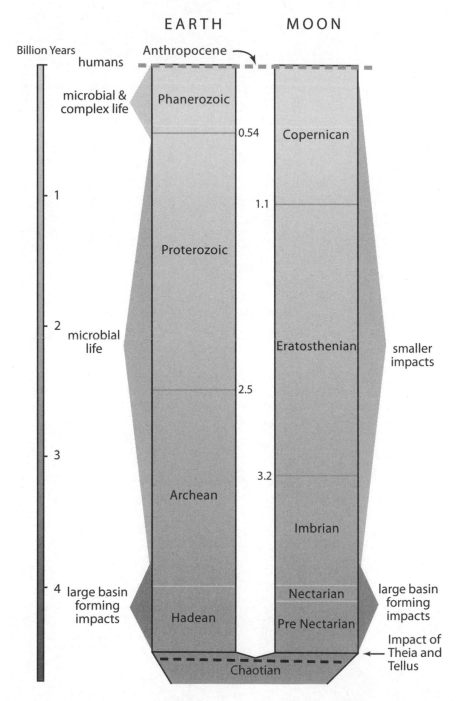

Following Grinspoon (2016), Goldblatt et al (2016) and authorial caprice

TRAJECTORIES

THE CHALLENGE IN REACHING IT IS NOT MERELY, OR EVEN MAINLY, one of distance. It is one of speed. Spacecraft, like planets and moons, are constantly falling, their trajectories shaped by the gravitational field of the Sun and of nearby smaller bodies. To get from one trajectory to another is a matter of changing velocity in the right direction by the right amount. The change in velocity required to get from one orbit to another is known as "delta-v". To get from the surface of the Earth to the surface of the Moon requires a delta-v of about 15 kilometres a second (km/s).

This needs to be applied in stages: every time a spacecraft changes its trajectory, it needs an extra dose of delta-v. The biggest requirement is the first. To get into a low orbit around the Earth requires a spacecraft to get up to a speed of 7.7km/s or so. In practice, to overcome various kinds of drag, you need about 9km/s.

From a near-circular orbit around the Earth the spacecraft then needs to get into a highly elliptical one, in which the perigee is still close to the Earth but the apogee is up by the Moon. That requires about 3km/s of delta-v. Once up at the Moon, the spacecraft then needs to change trajectory again to go into orbit around it. That requires another 1km/s.

If, once in lunar orbit, the spacecraft is to land, it needs another 2km/s to lose its orbital velocity and end up stationary on the surface.

Thanks to the lack of an atmosphere, you can get very close to the surface without that final step. Spacecraft have orbited the Moon at altitudes of 30km or so, and made one-off passes closer than that. Orbiting the Moon, though, raises other problems. The mass of the Moon's crust is not equally distributed, and concentrations of mass in the maria act like phantom reefs, running low-flying satellites aground from a distance unless their orbits are carefully designed so that the perturbing influences cancel each other out. Orbits higher than about 1,200km, on the other hand, are destabilized by the pull of the Earth.

If orbiting the Moon can be hard, coming back is easy. As a spacecraft goes out to the Moon, the Earth's gravity is pulling it back: it is, in effect, going uphill. On the return trip, as soon as it escapes the Moon's much weaker pull, the Earth's gravity does the rest. At the end of the fall you can get delta-v for free by turning your incoming velocity into the scalding heat of re-entry. It took Neil Armstrong and Buzz Aldrin less than 3km/s to get back to the Pacific from Tranquility Base, even allowing for a slight detour to pick up Mike Collins. As one of the characters in "Destination Moon" points out, a rocket as small and primitive as the V-2 would have been able to make the trip.

This demonstrates the way in which what matters in space is delta-v, not distance. To get from the Moon to the Earth requires only about a fifth of the delta-v that is needed to make the same journey in the opposite direction.

Another demonstration of this proposition is that the amount of delta-v it takes to get a spacecraft to the surface of the Moon can also get it to destinations much farther afield. "Near-Earth asteroids" (NEA) are only near inasmuch as they have orbits that very occasionally bring them moderately close to the Earth. At any given time a typical NEA will be 100 or 200 times as far away as the Moon. But in terms of delta-v, quite a lot of these asteroids are just as easy to reach as the Moon is; it is just a matter of getting on to the right trajectory and waiting. Indeed, it takes no more delta-v to reach the little moons of Mars than it does the great Moon of Earth (the surface of Mars is another matter). This means the Moon's closeness to the

Earth does not necessarily make it the most obvious place to go if one is keen on exploiting extraterrestrial resources.

Not all orbits have to be around a physical object. When a smaller object orbits a larger one, their gravitational fields combine to create a few places in empty space around which a spacecraft can orbit with minimum fuss. These are called "Lagrangian points" after Joseph-Louis Lagrange, the physicist who first systematically studied them. For the Earth and the Moon, there are two Lagrangian points at the same distance from the Earth as the Moon, one 60° ahead of it in its orbit, one 60° behind. These are called the L4 and L5 points. Two other points sit on the line between the Earth and the Moon. The L1 point sits a bit less than 60,000km above Sinus Medii in the middle of the Moon's nearside—that is, at a distance equivalent to about 30 times the radius of the Moon, or 15% of the way to the Earth. The L2 point is a little more than 60,000km above the middle of the farside; it sits almost directly above a crater named for Yuri Lipsky, the Soviet scientist who used data from Luna 3 to produce the first maps of the whole Moon. A spacecraft in a "halo orbit" around the L2 point, one that is perpendicular to the line joining Earth and Moon, can enjoy a view of the farside while also seeing the Earth over the limb of the Moon.

Uses for the opportunities provided by all of these Lagrangian points may eventually be found. Orbits around L2 already have. When China's Chang'e 4 landed in South Pole-Aitken in January 2019, it was able to receive instructions and return data only by means of Queqiao, a relay satellite in a halo orbit able to maintain radio links to the Earth and the lunar far side.

Queqiao means "bridge of magpies". Zhi Nu was a maiden from the sky who wove brocade out of the clouds for her father, the Jade Emperor. She fell in love with Niu Lang, a cowherd, who climbed into the sky to be with her. The emperor, disapproving of this, decreed that the lovers should live on opposite sides of the Milky Way, an impassable obstacle. But once a year, on the seventh day of the seventh month, magpies form a bridge across the barrier, allowing the two to commune. The tears of the briefly reunited lovers bring rain to the Earth.

- V -

REASONS

I T WAS WONDERFUL: PARENTS CAME TO PULL US TOWARDS THE televisions, which like the Moon were black and white and really, mostly, grey. Or they took us outside and pointed up at the sky. Or both. Sky Moon and screen Moon both there and both theirs and both ours, too; some large, strange change was being shared. Something grown-up that applied to children, too, and to what adults wish children to be.

Not all parents; not all of those of us then children. Not, for everyone, the most important thing or even a thing at all. For some an alienation, or a waste. But for many that encounter with history stayed bright, and for some it went deep. The excited adult pointing up promised a larger world that was yet to come, a world of the sky beyond the screen and the space in between for us to grow into.

We did not know that only 15 Saturn Vs had been built or that there were only nine left. We did not know that the last three Apollo missions were to be cancelled ("I just don't want to take the risk of a possible goof off," said President Nixon). We did not know that the USSR, having failed to build a successful Saturn-class super booster of its own, had no plans to put its own people on the Moon, or anywhere beyond low-Earth orbit, or that no one else was in a position to plan at all. We did not appreciate that, for all the excitement of 1969, the American public had never been overwhelmingly keen on Moon missions and that other priorities were looming larger, with price tags of their own. The Moonworld was there for everyone, a future history which, our parents seemed to promise us, would grow up as we did.

And then it didn't.

It is not that nothing happened. NASA built its space shuttles, and flew more than 100 missions with them, losing two orbiters and 14 souls in the process. America and Russia, in post–Cold War partnership with the European Space Agency, the Japanese and the Canadians, built an ambitious international space station, called it the International Space Station and have kept crews on it continuously for 18 years at the time of writing. For most of that time, rovers trundled across the deserts of Mars, geologizing as they went and the Cassini orbiter danced among Saturn's moons and rings. Communication satellites became a big business; space became vital to commerce, and to the military.

But it was never enough for the people sometimes called the "orphans of Apollo"—the children of the late 1960s for whom space remains an inspiration and a disappointment. The fact that Apollo had not seemed to come to anything may have disappointed many, but most of them don't regularly, or ever, give it a thought. For some, though, the curtailment of the space programme felt, as they came of age, like something between a bereavement and a betrayal. These were the people who asked, plaintively, "If we can put a man on the Moon . . . why can't we put a man on the Moon?" Or, some would add, a woman. They followed the space programme; they became scientists and engineers so they could work in it; they taught kids about it. Some of it inspired them, some of it thrilled them. None of it satisfied them.

They thought—they still think—that human progress and a human presence beyond the Earth were indistinguishable and good for everyone. They felt the desire to "put a dent in the universe" that Steve Jobs spoke of. But they had the unsettling knowledge that the dent they wanted to make had already once been made, and that the panel beaters of history had, with little fuss, removed more or less every sign of it.

They were, in an odd way, a mirror image of the curiously large group of people who denied that the landings had taken place in the first place, saying instead that they had been faked. Of course they hadn't been; of course they did. But did not the lack of a subsequent Space Age suggest, in a way, that they might as well not have done? And didn't that make them fakery of another kind? A dream that does not come true is not necessarily a lie. But, as Bruce Springsteen suggests in "The River", it could be something worse.

The dissonant loss of the future that their childhood and their beliefs told them was inevitable spurred them on. It also wore them down. All of them felt some bitterness, some resentment, some anger; some of them felt all three and more. They were angry at NASA, at Congress, at the military-industrial complex, at everyone responsible for the future in space coming so far short of what it should have been. They were angry at fools who would not listen, at cynics who listened only to scoff, even at other true believers who didn't adhere to quite the right version of the creed. At everyone who just didn't get it.

But if anyone was not getting it, it was the orphans. To a child, all new things are beginnings. Apollo had been an end: what the Space Race had built to, not what it would be built on. It retired what few plausible motives there had ever been for going to the Moon, both in America and everywhere else, and left behind no new reasons with which to replace them. America had successfully signalled to the world its technological pre-eminence. And though science had not been the primary purpose of the programme, advocacy by Shoemaker and others, and the enthusiasm that they engendered for their cause in the astronauts themselves, had made sure that the basic science had been done. What the Moon was made of; how it had come to be; what has shaped it since: all were basically answered.

Sure, there were types of lunar terrain that had not been walked on. The Earth-free sky of the farside had not been seen. Rocks produced by post-maria volcanism had yet to be sampled. But these were second-order concerns. Science had Mars to land its probes on, Jupiter and Saturn and their moons to fly past, Mercury to visit for the first time. It was headed out, not back—and it could do without astronauts. Space could be a frontier of the mind without the carefully wrapped still-fragile body.

Reaching for the Moon. Crying for the Moon. Walking on the Moon. Drawing down the Moon. They had all been images of fantasy, or frustration. And now they were again.

◑ ○ ◐

The hero of Robert Heinlein's "The Man Who Sold the Moon" (1950), Delos D. Harriman, is a tycoon determined to set the first Moon rocket on its way even if it costs him his whole fortune—and a lot of other people theirs as well.

> —Delos, why don't you give up? You've been singing this tune for years. Maybe some day men will get to the Moon, though I doubt it. In any case, you and I will never live to see it.

> —We won't see it if we sit on our fat behinds and don't do anything to make it happen. But we can make it happen.

As Harriman wheedles business partners, politicians and financiers into supporting his mania, he runs through a broad gamut of funding schemes and rationales. Some were the very ones which would be used by Apollo; in the decades that followed, all would be tried again and found wanting.

Ten years before John F. Kennedy's election campaign promised America "New Frontiers" Heinlein's Harriman offered inspiration

> —Tell them that this means new
> frontiers . . .

in the form of America's settler story of new lands and new horizons. The vision became part of Apollo, and its appeal persists to this day, felt intensely by a few, for most just one part of a more generalized exceptionalist belief that there is more to America than the mundane, some promise which inspires. The sentiment was strong enough to save America's human-spaceflight programme from the job-done post-Apollo cancellation that could easily have been its fate. In 1971 Cap Weinberger, then the director of the Office of Management and Budget, recommended in a memo that instead of cancelling the programme entirely, the nation should build the much-more-modest-than-Apollo space shuttle.

"America", Weinberger asserted, "should be able to afford something besides increased welfare, programs to repair our cities, or Appalachian relief and the like."

"I agree with Cap", President Richard Nixon wrote in pen at the top of the memo. Americans would continue to fly into space as others did not and would continue to see that as a bit of what made them special.

As had been the case with Apollo, the space shuttle was also sold as

> — . . . a shot in the arm for prosperity.

a path to profitable new technologies. In this respect it did not pay off. The sort of technology the world was interested in

> —Fast transportation will pay; it always has.

was changing: the ever-faster, ever-farther era that burst on it with the steam locomotives that rushed past the Bridgewater Foundry, and which heard its apotheosis in the thunder of the F-1 engine, was over. The taste for speed's genuinely modern pleasure was sated. No humans have ever gone faster than the crew of Apollo 13.

The shuttle's military applications were also part of its attraction. But they were limited to putting spy satellites into low Earth orbit. It wasn't as if there was anything farther out in space that people could any longer be scared

> —For years I've had a recurrent nightmare
> of waking up and seeing headlines that
> the Russians had landed on the Moon and
> declared the Lunar Soviet.

into racing to defend. Submarines offered a much better way of keeping nuclear weapons safe from sneak attacks than squirreling them away on the Moon. The Outer Space Treaty negotiated at the United Nations in 1967 was able to enshrine the idea that space must only be put to peaceful uses largely because the two powers with the capacity to make non-peaceful use of it could see no real mileage in doing so.

Despite the impact the first broadcasts from the Moon had on adults and children around the world, it was never going to be a useful communications relay, because it could broadcast to any given spot

> —Did it ever occur to you that there
> is absolutely no way to interfere with a
> telecast from the Moon and that boards of
> censorship on Earth won't have jurisdiction?

on the spinning Earth for fewer than 12 hours in every 24. As Arthur C. Clarke had pointed out, satellites in geostationary orbits were far superior. Witness the fact that, without them, NASA would not have been able to relay the messages picked up from its receivers all around the world back to Houston—and Walter Cronkite—regardless of where the Moon was in the sky.

The Moon's minerals

—A thousand acres at a dollar an acre.
Who's going to turn down a bargain like
that? Particularly after the rumour gets
around that the Moon is believed to
be loaded with uranium?

—Is it?

—How should I know?

were, in the 1970s, of academic interest only. The Apollo samples held
no remotely attractive ores. The diamonds with which science fiction had
been littering its surface since the 1930s

—Your geologists all agree that diamonds
result from volcanic action. What do you
think we will find there?

He dropped a large photograph of the
Moon on the Dutchman's desk. The
diamond merchant looked impassively
at the pictured planet, pockmarked
by a thousand giant craters.

were nowhere to be seen. The pressures needed to form diamonds are
found only in the deepest portions of the Moon's mantle, a zone from
which there is probably no escape to the surface.

There are other ways a venture could raise money. But advertising,

—A few days ago a man came to me—
you'll pardon me if I don't mention names?
You can figure it out. Anyhow, this man
represented a client who wanted to buy
the advertising concession for the Moon.
He knew we weren't sure of success; but
he said his client would take the risk.

product placement,

When the Moon-ship *Pioneer* climbs
skyward on a ladder of flame, twenty-seven

essential devices in her "innards" will be
powered by especially-engineered
DELTA batteries . . .

raiding children's piggy banks

—I want an angle to squeeze dimes
out of the school kids, too. Forty
million school kids at a dime a head
is $4,000,000.00—we can use that.

—Why stop at a dime? . . .

and philanthropy

—It's never milked dry, as long as there
are rich men around who would rather
make gifts than pay taxes.

were not going to do it. Nor, for that matter, would philately.

—George, you collect stamps, don't you?

—Yes.

—How much would a cover be worth which
had been to the Moon and been
cancelled there?

—Huh? But you couldn't, you know.

—I think we could get our Moon-ship
declared a legal post office sub-station
without too much trouble.

The crew of Apollo 15 actually tried this one, taking some stamped
commemorative covers that had somehow not made it onto NASA's
official manifest down to the Moon's surface and later providing them
to German collectors. The rewards were not that great. And none of
the three men ever travelled into space with NASA again.

In the end, though, whatever reasons and inducements Harriman
could offer his friends and marks for going to the Moon were not the

real point. As Harriman, and Heinlein, knew from the first words of their story

> —You've got to be a believer!

You. Not everyone. You. The vision of an expansion beyond the Earth and into the future sunrise

> To the east the ship towered skyward,
> her slender pyramid sharp black
> against the full Moon.

is a curiously personal thing. Those with the vision believe in its universal relevance; they are mostly, I think, sincere in what they say about human destiny and the like. But they also believe this universal good can and will act specifically through them, and people like them. In "The Challenge of the Spaceship", a paper that he delivered to his colleagues in the British Interplanetary Society in 1946, Arthur C. Clarke talks of explorers, musicians and mathematicians as people who do what they do not for any practical or impractical reason, but because they must:

> [And] so, if we will be honest with ourselves, it is with us. Any "reasons" we may give for wanting to cross space are afterthoughts, excuses tacked on because we feel that we ought, rationally, to have them. They are true, but superfluous—except for the practical value they may have when we try to enlist the support of those who may not share our particular enthusiasm for astronautics yet can appreciate the benefits it may bring, and the repercussions these will have upon the causes for which they, too, feel deeply.

The orphans of Apollo mixed Clarke's particular enthusiasm and Harriman's compulsion to believe with the frustration of being not tantalized by a fancy of the future but robbed of a fact of the present. They thought that the world wanted the Moon and what it stood for. In fact, as Clarke saw would be the case, powerful men had wanted it instead for the repercussions they believed it could have on other causes they felt more dearly.

● ○ ◐

Now, AFTER ALMOST 50 YEARS OF HURT, BUT NEVER STOPPING dreaming, the orphans' wait is almost over. A flotilla of robotic payloads will be beaching up on the lunar surface in the next five or so years, some from established spacefaring powers like China, India and America, some from newcomers, such as Israel and Canada. Some will be paid for as business investments, and some as philanthropy, instead of by governments, and some by money from all those sources. Some will get there under their own steam; some will pay for a ride on another company's, or country's, bus. Some will be given their rides for free.

In part this is because the technology of robotic spaceflight has got better, and cheaper, and the means of private individuals interested in such technologies have grown larger. In part it is because new questions arose in the 1990s—specifically, the nature of those volatiles at the Moon's poles—that people want to answer both for scientific reasons and for practical ones. And in part it is because it looks increasingly likely that humans will return to the Moon reasonably soon, not simply to visit but perhaps to stay; there is talk of moonbases again, even of "Moon villages." The robots are their scouts—some of them in a practical sense, all of them in a symbolic one.

As well as new technology and the new sources of money, though, there are also new reasons. One is descended from the Cold War that launched the Space Race and the lunar landscapes it revealed: the issue which has come to be known as "existential threat".

One of the reasons people in the 19th century and the first half of the 20th were reluctant to believe that the Moon's craters were formed by impacts was the disturbing corollary that if the Moon was thus battered, the Earth must be, too. Understanding of the distinctive impact signatures that Shoemaker first documented at Meteor Crater proved this correct. In 1961 he and his colleagues showed the

Nördlinger Ries in Germany to be a far larger impact structure, a crater 24km across. A couple of years later the impact origin of the Carswell structure in Saskatchewan was recognised. This one was 40km across—almost half the size of Tycho, and roughly the same age.

Geologists, the uniformitarian custodians of Earth's history, at first saw these past impacts as irrelevant to their greater concerns and inimical to their distrust of the catastrophic. Some science fiction writers paid more attention. In "Lucifer's Hammer" (1977), Larry Niven and Jerry Pournelle provided a physically plausible and politically troubling account of a comet impact and its apocalyptic effects. In 1980 Walter Alvarez, a palaeontologist at the University of California; Luis Alvarez, a physicist and his father; and Frank Asaro and Helen Vaughn Michel, two chemists, revealed evidence that a large impact had brought about the end of the Cretaceous Period, and with it the extinction of the dinosaurs, by scouring North America with tsunamis and debris and throwing enough dust into the atmosphere to deprive the whole planet of sunlight. "'Lucifer's Hammer' killed the dinosaurs", Luis Alvarez told one of his audiences.

By this point Shoemaker and his wife, Carolyn, as well as a few other researchers, were using telescopes to scan the skies for Earth-threatening asteroids. Worrying about such things was still seen as absurd in policy circles, but not among space enthusiasts, for whom it underlined two previously marginal reasons for going into space.

One was to find any asteroids or comets that threatened Earth and deflect them. Rick Delanty, an Apollo-era astronaut in "Lucifer's Hammer", is clear and anguished on the question: "In ten more years we'd have been able to push the damned thing out of our way!" At science fiction conventions people began to be seen in T-shirts that said, "The dinosaurs died out because they didn't have a space program"—a line attributed to Mr Niven, along with the rider "And if we become extinct because we didn't have a space program it will serve us right!"

The other reason was a deeper, broader one, built on an urge not to deflect but to flee. The insight that Kevin Zahnle and Norm Sleep had into life during the heavy bombardment—that when the shit goes down, it is better to be thrown off the world on a meteorite than to boil in situ with the oceans—could apply to humans in the Anthropocene just as it applied to microbes in the Hadean. Being off the planet provides a refuge when things on the planet can't be survived. As Clarke put it in his first published novel, "Prelude to Space" (1948): "The Earth is just too small and fragile a basket for the human race to keep all its eggs in."[*]

The thought is another of the ways that space travel and super-weapons form a science-fictional double act, with the rockets which might prosecute a nuclear war reinterpreted as the means of escaping it. As "Prelude to Space" has it, "Atomic power makes interplanetary travel not just possible but imperative". In Ray Bradbury's "The Martian Chronicles" (1950), American colonists look back on their Earth engulfed in nuclear flame, the first characters in science fiction, I believe, so to do.

Side-stepping war this way might work—to the extent that, survivor guilt being what it is, it could work at all—if the colonists are overwhelmingly from one side of the conflict. But if the people of Mars, or the Moon, are as divided as the governments of Earth, the war would seem likely to spread. In Michael Swanwick's "Griffin's Egg" (1992), the workers on the Moon's surface who watch in shock as pinpoint-clear nuclear blasts pepper the face of the Earth above them return to a moonbase driven mad by neurotoxic sabotage. In Ben Bova's "Millennium" (1976), the inhabitants of adjacent American and Soviet moonbases built during a long-gone era of détente watch their home world move inexorably towards war. Chet Kinsman, the American base commander, suggests to his counterpart Piotr Leonov that they unite, declare independence and sit out the war as only those 380,000km away have the privilege to do.

> "Do you seriously believe," [Leonov] asked slowly, without turning back to face Kinsman, "that any of us could watch our homelands

---

[*] "Lucifer's Hammer" ascribes this line in a slightly different form to Heinlein.

being destroyed without going mad? Do you honestly believe that their war will not destroy us, too?"

Forcing his voice to stay calm as he walked to stand beside his friend, Kinsman answered, "We could get through it without fighting. If we tried."

The Russian's voice was infinitely sad. "No, old friend. I might trust you and you might trust me, but to expect nearly a thousand Russians and Americans to trust each other while they watch their families being killed—never."

Kinsman wanted to scream. Instead, he heard himself whisper, "But Pete, what can we do?"

"Nothing. The world will end. The millennium is rushing upon us."[*]

Sitting out a nuclear war on the Moon had a ghastly fascination. As a rationale for going to the Moon in the first place, though, it was terrible. Surely better to put one's resources into stopping the war than into building a sad, squalid bunker in the sky?

Inhuman threats to life on Earth, such as impacts on a dinosaur-extinguishing scale, seemed to offer a less suspect motive for a planetary exit strategy. Although war might always be better averted than fled from, a natural precariousness to life on Earth really might justify expansion. Insights from the Moon into the Earth's battered history reinvigorated this idea, but it was not new. Oriana Fallaci's book about astronauts and Apollo, "If the Sun Dies" (1965), is framed as a conversation between Fallaci and her father, who sees all that life needs, and all that life is—air, water, growth—as being always already available on Earth: why then venture into asphyxiating, lifeless space? Fallaci replies by quoting Ray Bradbury—not in his fictional mode, but in his prophetic one:

---

[*] In Stanslaw Lem's "Peace on Earth" the idea of the Earth's battles spreading inexorably to the Moon is taken to its logical—which is to say, absurd—conclusion: robots on the Moon fight proxy wars so that the Earth can live in peace.

For the same reason that makes us bring children into the world.

Because we're afraid of death and darkness, and because we want to see our image reflected and perpetuated to immortality.

We don't want to die, but death is there, and because it's there we give birth to children who'll give birth to other children and so on to infinity.

And this way we are handed down to eternity.

Don't let us forget this: that the Earth can die, explode, the Sun can go out, will go out.

And if the Sun dies, if the Earth dies, if our race dies, then so will everything die that we have done up to that moment.

Homer will die, Michelangelo will die, Galileo, Leonardo, Shakespeare, Einstein will die, all those will die who now are not dead because we are alive, we are thinking of them, we are carrying them within us.

And then every single thing, every memory, will hurtle down into the void with us.

So let us save them, let us save ourselves. Let us prepare ourselves to escape, to continue life and rebuild our cities on other planets: we shall not long be of this Earth!

The words resonate and inspire: they sound like an evocation of the loftiness of the human spirit. At the same time they feel as desperate as they do heroic. And they have more in common with some of the intellectual criticism of the Apollo programme Fallaci celebrates than either side would have accepted at the time.

That criticism, too, was motivated by a fear of the end of the world. It just put the end in the present rather than in the future, seeing it as a piece with the thinking that had brought about the space programme in the first place. As Hannah Arendt argues in "The Human Condition" (1958), to wish to go beyond the Earth was to break something fundamental about what it is to be a human in the world. Influenced, as a student and enemy of totalitarianism would have to be, by the fact that Sputnik had been a Soviet achievement, Arendt, like Fallaci's father, saw a form of annihilation in such attempts at technological transcendence. The human condition was to be rooted in a world of life and death, to be born and nourished by it, to die in it. Space travel was thus in itself the end of the world. Martin Heidegger, her by then long-repudiated teacher and lover, expressed similar views in a 1966 interview published after his death a decade later:

> Technology tears men loose from the earth and uproots them. I do not know whether you were frightened, but I at any rate was frightened when I saw pictures coming from the moon to the earth. We don't need any atom bomb. The uprooting of man has already taken place. The only thing we have left is purely technological relationships. This is no longer the earth on which man lives.*

These, too, are feelings that have a home in science fiction. But look for them in the interior catastrophes of J. G. Ballard, not the cosmic crashes of Niven and Pournelle.

● ○ ◐

THE TECHNOLOGICAL WORLD PICTURE THAT HEIDEGGER DEPLORED was one in which the orphans of Apollo revelled, and its expansion became the most rehearsed of their reasons for getting back to space, and the Moon: industry and private enterprise.

The first influential outline of a space programme driven by the fulfilment of an economic and global need rather than reasons of state or destiny came from Gerard K. O'Neill, an idealistic Princeton physics

---

* It is interesting that these remarks pre-date the Apollo 8 "Earthrise" picture by two years. Unless Heidegger was hallucinating, this means that he was one of the few to see, and realize the significance of, an earlier and grainier picture of the Earth over the Moon sent back by one of the Lunar Orbiter spacecraft. Perhaps someone made sure that, as the person who had introduced the term "world picture" to philosophy, Heidegger should get to see it. ●

professor. Having seen Apollo fall back to Earth, he argued that long-term expansion into space needed to provide continuous benefits, year in year out, rather than a big rush of pride. This led to a scheme which offered to solve what, in the mid-1970s, seemed one of America's, and the world's, biggest problems: the energy crisis.

O'Neill imagined vast arrays of solar panels in geosynchronous orbits, immune to night and cloud, soaking up the harsh unfiltered 192-proof sunlight of space 24/7. They would turn the electricity into microwaves and send it down to receivers on the surface of the Earth using wavelengths that the atmosphere would not absorb. From there, the power would flow into national grids. These solar power satellites would provide the world's energy needs without air pollution, without nuclear meltdowns, without dependency on OPEC, without—though they were, at that point, hardly an issue—fossil fuel emissions.[*]

Launching such huge structures from the Earth was not remotely feasible. O'Neill reckoned that a power satellite in geosynchronous or-bit capable of providing a gigawatt of electricity—a thousand mega-watts, the output of a large conventional power station—might have a mass of 16,000 tonnes. The then-not-yet-built space shuttle was to have a cargo capacity of less than 30 tonnes, and even that mite would be lifted only to low orbit. The only way to build things as big as power satellites was to get most of the material from the Moon.

O'Neill's plans were to use bulldozers to strip-mine the regolith, processing the dust and rubble for metals and silicates of the sort that solar panels can be made from. These raw materials would be flung into space using a "mass driver" not unlike the "maglev" trains which are both supported and accelerated by electromagnetic fields, floating frictionless over their tracks. Such a railway, laid out in a straight line across the lunar surface, could accelerate its cargoes to orbital speeds.

---

[*] In Harriman's search for reasons for spaceflight Heinlein had of course included en-ergy—but rather than safe, clean energy, he had imagined space as a place to put danger-ous energy, in the form of nuclear power plants permanently on the brink of explosion.

The trajectories these materials were thrown onto would take them to the Earth-Moon system's L5 point, 60° behind the Moon in its orbit. There a workforce much larger than that required for the Moon-mining operation would use them to manufacture solar power satellites. That workforce would live in "Islands" of their own creation, also built from lunar raw materials. O'Neill imagined one such structure—he called it "Island Three"—as a hollow cylinder kilometres long, spinning on its axis so that the workers living within would enjoy a centrifugal force that emulated gravity.* Sunlight would be let in through long windows; the neighbours would live overhead.

This vision, published under the title "The High Frontier" (1976), proved to have a wide and eclectic appeal. Tech-heads liked it; hippies liked it, too; so did the ecology minded. Stewart Brand, a magnificent Californian impresario of ideas who had campaigned for NASA to release pictures of the whole Earth from space when they were not available, took up the cause in his publications *Whole Earth Catalog* and its spin-off *CoEvolution Quarterly*. The same publications had also, not coincidentally, been the venue for some of the first serious discussions of James Lovelock's Gaia hypothesis. Gaia was part of the anti-Copernican shift back to the Earth driven by how lifeless everywhere else looked; it celebrated the specialness of such a living system. O'Neill's spin on this was to suggest that the way to correct the deficit of worlds like the Earth in the sky was to build them from scratch in inside-out miniature.

Rick Delanty, the frustrated astronaut in "Lucifer's Hammer", harangued guests at his Houston barbeques about O'Neill's ideas. Jerry Brown, the governor of California, was interested in them, too. California is a land of dreamers, of environmentalists and of aerospace companies; something which appealed to all of the above was worth looking into. O'Neill inspired the first new grass-roots activist community devoted to space since the 1930s of the Verein für Raumschiffahrt, the British Interplanetary Society and the American

---

* Anyone disturbed by this use of language is referred to xkcd.com/123.

Rocket Society.* At the movement's centre was a new organisation called the L5 Society supported by Heinlein himself.

Part of the appeal of the O'Neill programme was its unified response to the two environmental concerns that were coming together in the 1970s, most famously in "The Limits to Growth" (1972), a report by the self-appointed "Club of Rome", and at the first UN environment conference, held in Stockholm the same year. Holding the post-Apollo icon of the whole Earth to its heart, this new environmentalism combined the fear that humans were now damaging the environment on a global scale with the fear that they would deplete the whole world of the raw materials that they needed. In a doom-laden decade, it provided a soft apocalypse hardly any less scary, and infinitely more widely worried about, than the sudden sharp impact of an asteroid.

The era's most gloomy environmental concerns were summed up in an identity formulated by Paul Ehrlich and John Holdren, academics at Stanford:

$$\text{Impact} = \text{Population} \times \text{Affluence} \times \text{Technology}$$

The more the world had of any of the three things on the right-hand side of this "IPAT" formula, the argument went, the more adverse the impact on the environment. But O'Neill and his followers felt that, like a reflection, or a pocket turned inside out, or a world in an unworld's sky, space technology could invert this counsel of despair by replacing technologies that had an impact on the Earth and drained its resources with technologies that did not. The IPAT assumption was that technology multiplied the impact of population and affluence. The L5-ers claimed that space-based technology decreased it. $I = PA/T$. If this denominating T is big enough, impacts can shrink even as more people get more affluent.

---

* The ARS had by this stage been rolled into the American Institute of Aeronautics and Astronautics, a thoroughly professionalized institution; the BIS continued in its quirky, part-time way, as it does to this day and will, I hope, continue to do for centuries to come.

Believers in this inversion promised endless spacey cake and end-less Earthly eating. They saw all sorts of heavy industry migrating to orbit, taking advantage of the unlimited energy, freely available vacuum and sophisticated manufacturing techniques only possible in the microgravity available in freefall. They talked of foamed metals lighter than candyfloss and tougher than steel; single-crystal whiskers stronger than hawsers; composites that mixed substances immiscible on Earth. This orbital industry would feed on raw materials from far-ther away, not just from the Moon but also from mines among the asteroids, where vast mirrors would smelt metals for the factories in orbit around the Earth below. Any pollution would be swept away by the solar wind, blowing every vapour and residue in its path out to the edge of interstellar space even more effectively than tides cleanse an estuary. Space as workshop; space as foundry; space as provider of sanitation: James Nasmyth would have loved it.

And this High Frontier would never close. It would just get higher and higher. Boosters like Pournelle argued that it would allow hu-mans not just to survive but also to "survive with style". Cosmism as capitalist self-improvement. More and more Moonshots: ever fewer have-nots.

Not that the Moon was, in itself, the point. True to its modern na-ture, the Moon was somewhat peripheral to such plans—just a source of raw materials. The would-be settlers of the High Frontier were by and large not that interested in the Moon, per se, with its already-visited deserts and close, confining horizons. The action would be at L5 and its purpose-built Islands.[*] It was they which best embodied America's love of starting over, they which made concrete the po-tential Thomas Paine spoke of when he said, "We have the power to

---

[*] As the science fiction author Ben Bova pointed out in his novel "Colony" (1978), a sequel to the rather better "Millennium" discussed above, the L5 movement turned its back on the Moon with its very name. If you cared about the Moon, you would put your colonies at L4, a point with all the same gravitational advantages but which also offers a particularly beautiful view of the Moon dominated by the great bull's-eye basin of Mare Orientale.

begin the world over again", they which offered the possibility of a second creation. The Moon was just the debris of the first.

Ideology aside, the O'Neill scheme had a practical disadvantage. It wasn't. Not practical at all. Even if energy prices had stayed at their 1970s crisis level, and even supposing, as O'Neill did, that tens of tonnes of equipment on the Moon could send thousands of tonnes of raw material to L5, just getting tens of tonnes of equipment to the Moon on a regular basis was far beyond the capacity of the space-shuttle fleet. The enthusiasts claimed that a much more efficient launch was possible. But the very existence of the shuttle fleet showed that government was not going to develop it. The enthusiasts claimed that lunar mines and solar power satellites would pay off almost as quickly as the multi-decade investments made in Earthly mines and nuclear power plants. Private capital remained spectacularly uninterested.

● ○ ◐

IN THE 1980S A WAY AROUND THIS IMPASSE WAS SUGGESTED. What if the Moon could be mined not for bulk materials used in far-out space colonies but for something of great value right here on Earth. If the Moon produced something worth tens of millions of dollars a tonne, it might be worth industrialising for that alone. The candidate wonder substance was helium-3.

Not all the solar wind that blows out from the Sun gets to interstellar space; some hits the surfaces of planets, moons and asteroids that lack the magnetic fields needed to deflect it. Some is thus absorbed by the lunar regolith. That wind contains helium-3, an isotope which is in some ways an ideal fuel for fusion reactors and is vanishingly rare on Earth.

Nuclear fusion produces energy by melding very light atomic nuclei into slightly heavier ones. In space it powers the stars. On Earth it powers hydrogen bombs. In theory—and it is a theory that has now enchanted several generations of physicists—fusion also offers an appealing alternative to nuclear fission as an almost limitless source of electricity which neither requires an infrastructure which can also enable nuclear weapons nor produces nuclear waste. There is a huge international programme aimed at building such a reactor in the South of France.

That reactor, ITER, will react deuterium, a stable isotope of hydrogen easily separated out of seawater, with tritium, a short-lived isotope of hydrogen that would have to be manufactured for the purpose. There are practical reasons for this fuel mix, but it is not ideal. As well as being radioactive, tritium is also widely used, if not strictly speaking necessary, in nuclear weaponry. And tritium-fuelled reactors would give off enough neutrons to turn some of a reactor's parts into low-level radioactive waste in need of eventual disposal.

Burning deuterium with helium-3 instead of tritium would avoid both those problems. Helium-3 is neither radioactive nor bomb-relevant. And fusing it with deuterium produces protons, not neutrons. Those protons, which carry an electric charge, can be used and disposed of without making anything else radioactive. The promise of helium-3 is thus the same as the promise of solar-power satellites: clean energy. But if you have the right reactor, you would need just 100kg of helium-3 a year to provide the same gigawatt of power as one of O'Neill's 16,000-tonne solar power satellites. It would take only a few hundred tonnes of the stuff a year to provide all the Earth's current electricity needs.

The idea of helium-3 mining was, understandably, taken up enthusiastically by L5-ers and science fiction writers. It is the basis of, among other things, Ian McDonald's "New Moon" (2016) and "Wolf Moon" (2018), and Duncan Jones's film "Moon" (2012). Harrison Schmitt, the geologist who went to the Moon in *Challenger*, Apollo 17's LM, is quite the devotee.[*] But like O'Neill's L5ism—indeed, rather more so—this idea, too, is profoundly impractical.

You would have to process tens of millions of tonnes of lunar regolith to get that 100kg of helium-3, an undertaking not that much

---

[*] Though given that Mr Schmitt also has form as a climate-change denier, I am not quite sure what he thinks the pressing need for fusion power actually is.

more manageable than flinging thousands of tonnes of the stuff out into space to be smelted and turned into satellites. And the drawbacks that helium-3 seeks to remedy are not the problems that are delaying the development of fusion power. The problems people actually working on fusion worry about are those involved in getting the technology to the stage where it can plausibly generate power at all. They have been working on this for decades; they foresee decades more work to come.

And that is for a tritium reactor. Burning helium-3 is far harder. But it is not all that much better. It is absurd to think that if tritium reactors become a reality, people will look at their relatively minor drawbacks and promptly decide to start work on much more challenging reactors that require moondust mines for their raw materials. The Earth very much needs many forms of non-fossil-fuel energy. But helium-3 only looks like a useful part of that portfolio if you start from the position of requiring an answer to make use of the Moon. That is not most people's starting point.

What is more, even if you do take the Moon as your clean-energy starting point, you might not light on helium-3 as your answer, or on solar power satellites, either. Dennis Wingo, an entrepreneurial orphan of Apollo who left the software business to work on space technologies, points out that the Moon could be a rich source of platinum-group metals. This is because about 3% of the asteroids that have pummelled it for the past four billion years are made of metal, not rock. Even smallish fragments left by such impacts would be worth billions, if not trillions, on the Earth's metal markets.

Mr Wingo is not ignorant of the law of supply and demand. He knows that if a lunar-mining concern were to offer the realistic prospect of huge new supplies of platinum, prices would plummet accordingly. But he also understands that cheap things can be more valuable than expensive ones. As an example, he cites aluminium, which when first produced in the early 19th century was more expensive than gold and mostly used simply as a way of showing off; Napoleon III had a set of aluminium cutlery which was set at the place of honoured

REASONS   191

dinner guests. In the following decades the metal's engineering possibilities became clearer, but its price remained a problem. Witness
the discussion which follows Barbicane's suggestion that it be used
to fashion the space capsule in Jules Verne's "From the Earth to the
Moon":

> "Aluminium?" cried his three colleagues in chorus.
>
> "Unquestionably, my friends. This valuable metal possesses the
> whiteness of silver, the indestructibility of gold, the tenacity of iron,
> the fusibility of copper, the lightness of glass. It is easily wrought,
> is very widely distributed, forming the base of most of the rocks,
> is three times lighter than iron, and seems to have been created
> for the express purpose of furnishing us with the material for our
> projectile."
>
> "But, my dear president," said the major, "is not the cost price
> of aluminium extremely high?"
>
> "It was so at its first discovery, but it has fallen now to nine
> dollars a pound."
>
> "But still, nine dollars a pound!" replied the major, who was not
> willing readily to give in; "even that is an enormous price."
>
> "Undoubtedly, my dear major; but not beyond our reach."

The price was to fall a fair bit further;[*] by the time Apollo's spacecraft were made of aluminium, as Barbicane had, in effect, advised, so
was a great deal of the rest of the modern world. The metal had become
cheap; indispensable to various industries, it was also very valuable.
Wingo imagines that a similar fall in prices for platinum and related
metals would allow it to become similarly valuable, specifically because
it would make hydrogen fuel cells far cheaper, thus—again—providing
a cleaner, more affordable energy infrastructure. I somewhat doubt this.
But it still feels more plausible than the helium-3 tarradiddle.

---

[*] This morning aluminium costs less than $1 per pound on the London Metals Ex-
change. Accounting for inflation, that makes it about 140 times cheaper than it was in
Barbicane's day.

MOST OF TODAY'S MOON-MINING ADVOCATES, THOUGH, CONCEN-
trate neither on metals nor on helium but on the ice and other volatiles
in the permanent shadows at the poles. Their exploitation might provide settlers with a reasonably plentiful local source of water as well as
some of the carbon, hydrogen and nitrogen that life needs in moderate
abundance but of which moonrocks offer more or less none.

By raising the possibility that a settlement might have the wherewithal to provide its own water, volatiles on the Moon reduce the practical burden that any other reasons for returning to it might need to
bear. And they might also provide a way to defray some of the costs.
Getting a tonne of payload from the Moon to low Earth orbit takes
a lot less fuel than getting it there from the Earth. So, if people doing
things in low Earth orbit need fuel and water, it might be cheaper to
send it to them from the Moon than from the Earth.

Like the platinum-group-metals story, though, this highlights
another issue about lunar resources. They may have competition. The
helium-3, Mr Wingo's metals and the polar volatiles all come from
elsewhere; the helium, true to its name, from the Sun, the metals and
volatiles from asteroids, comets and some water-rich inbetweenies.
Why not go directly to the source? An icy carbon-rich asteroid might
be a more amenable source of fuel for satellites orbiting the Earth
than the grubby ice caplets at the Moon's poles. In terms of delta-v, if
not travel time, it could also be closer. And though one asteroid could
not compete with all the Moon's ice, there are many asteroids. Similarly, a metal-rich asteroid might be a better source of platinum-group
minerals—though the Moon, having accumulated the debris from
such asteroids for billions of years, may have some particularly choice
nuggets secreted about its person.

For some space enthusiasts this doesn't matter at all: if asteroids
deliver the goods and the Moon doesn't, then go mine the asteroids.
For those imprinted on the Moon itself, for those who look up at its
face and know that it is that world in reflection, not space in general,

that they want, asteroid mining carries the threat of lunar marginalization, even irrelevance.

It is not the only such threat. To many for whom an interest in, even devotion to, space is mainly driven by a love of science, the Moon is not all that appealing, at least, not compared with Mars. The same also applies to those who see space as a way to signal something through a grand and unprecedented achievement.

Going to Mars is a far greater engineering challenge than the Return to the Moon. The trips out and back take months, not days, and can only be made at all when the two planets are appropriately aligned; this means plausible mission architectures either last for years or provide very little time on the surface. A life-support system that can keep working for years with no resupply flights is a challenge no one has yet taken on. And the longer a mission is away from Earth, the more likely it becomes for chains of independent, rare events to lead to unforeseen problems and dangers.

At the same time, Mars is the only planetary surface other than the Moon's and the Earth's that is plausibly accessible using today's technology. The delta-v needed for Mercury or the moons of Jupiter is far too high, even before you start to worry about the savage sunlight bombarding the former or the vicious radiation belts around the latter. Venus is like the depths of the ocean would be if the pressure were greater and the water hot enough to melt lead.

And Mars is fascinating in ways the Moon is not. It has an atmosphere that moves its sand and dust round; in the past it had flows of ice and water capable of similar erosional and sedimentary services, as well as vast volcanoes. It thus has a rich geological history, beautifully revealed in recent years by the rovers on its surface. It may have a biological history, too. It could well have been the abode of life in the past; it might conceivably have some simple organisms deep within its crust today. In the intermingled worlds of science fiction and scientific speculation it is accepted as easily the best target for the intentional, directed climate change known as terraforming. Its atmosphere could be thickened, its climate warmed;

it could support surface water, even plants. The red planet could have its own red edge.

This worldliness-in-waiting gives Mars a mystique. Robert Zubrin, an aerospace engineer who in 1998 founded the Mars Society, sees the settlement that the society advocates as a way—perhaps the only way—to regain a cultural vigour he thinks was lost with the closing of the American frontier at the end of the 19th century. Elon Musk, founder of SpaceX and, as I write, probably the world's most talked-about entrepreneur, sees Mars as a hedge against existential all-eggs-in-the-same-basket disasters. In a messy mix of cosmic compassion and messianic self-belief, Mr Musk is set on making humanity a multiplanetary species, and Mars—eventually, a terraformed Mars— is the first step on that road.

Its mixture of mystique, new challenges and science has ensured that whenever the US government sets out long-term space plans, human feet on the sands of Mars are always in the mix. They were there in the Space Exploration Initiative proposed by George H. W. Bush in 1989, and in the Vision for Space Exploration his son promulgated in 2004. So, too, was the Moon. But as a long-term goal, Mars has tended to overshadow it. In many minds, the two goals have become, to some extent, competitors. Some see it as a friendly fight; some treat it as a fierce one.

Like many competitions, it is easily understood by those who live it, yet hard to grasp from the outside. To set Mars against the Moon is, to various people, and to various extents, setting science against commerce; new beginnings against continued growth; symbolism against action; arcadia against industry; later against sooner; establishment against underdog; pointy-heads against working stiffs. It is thus to some extent, like so much in America, left against right. Each opposition can be questioned, and none is fundamental. But in general Mars appeals to dreamers, to those with a deep commitment to that as yet undone, to seekers after deep but impractical scientific knowledge, specifically the astrobiological knowledge of life beyond the Earth.

Though the Moon has astrobiological charms, too, in its early Earth rocks and records of bombardment, they are for most purposes less inspiring. But it is better placed to satisfy the desire that there should be a continued presence beyond the Earth. Moon boosters stress far-fetched economic returns such as those dependent on helium-3 because they see them as providing a strong foundation for permanent expansion; if space becomes a place to make money, it will not again be abandoned. In this, they are true heirs of Harriman. And what's wrong with making money? There are quite a few space activists who have become entrepreneurs, and though they are seeking an impact beyond financial returns, such returns would still be welcome, not just as a way of encouraging further investment but also because what good capitalist doesn't want more money rather than less?

Mars makes the orphans of Apollo on the lunocentric side of the argument worry that their childhood trauma will be repeated—flags and footsteps will be duly left in the red dust, the astronauts will come home and nothing else will change. There is, after all, no commercial reason to go to Mars.* They also worry, with reason and experience on their side, that nothing suits a political system better than a feel-good goal beyond the terms of office of all concerned. If you don't really want to do much in space, then saying that what you want is to go to Mars sometime in the late 2030s is a pretty good tactic; it keeps NASA and its associated contractors ticking over but doesn't require it to get more money. Nor does it expose you to association with failure. Nixon's worry that NASA might goof off has surely been shared by some of his successors.

● ○ ◐

THE ORPHANS' LAST REASON FOR SPACE IS BOTH THE MOST TRIVIAL and the deepest: tourism. They all want people to go to space. Many

---

* In his novel "The Secret of Life" (2001), Paul McAuley does manage to imagine something both Martian and economically important, but it is not the sort of thing one could reasonably set out in search of.

of them want to go to space themselves. So, going to space just in order to go to space might be a business in itself. By the 1990s, millionaires were getting interested in the possibility of trips to *Mir*, the once-Soviet, then Russian space station. "Orphans of Apollo" (2008), a documentary by Michael Potter which first brought that phrase to an audience, tells the story of a failed attempt to take *Mir* into private ownership in part as a way of making this service available. That did not happen—but from 2001 to 2009 various rich men paid to be taken up to the International Space Station in Russian capsules.[*]

In the next couple of years suborbital space tourism—flights that get higher than 80km, but without anything like the delta-v needed for orbit—look likely to begin. It seems there are hundreds of people, maybe thousands, willing to pay $200,000 or so for flights which will reveal the Earth in its planetary pomp, the curvature of its horizon clear, the sky above it black, the clouds far, far below, with a few minutes of weightlessness thrown in. There are plans afoot for private journeys into orbit and new hotel facilities on the space station to welcome them.

To some, this might look a little unseemly. If the sustained expansion of a human presence into space is the inauguration of a new phase of history, or the discharging of a sacred trust, or an imperative for the survival of the species, is it not a little tawdry to reduce it to thrill-seeking by the rich? Gagarin challenged the cosmos; Apollo dealt in nobility and the coordination of the world's greatest economy; what is noble about just buying a ticket, having an experience and coming home?

Against this distaste, though, are a few other considerations. The first is practicality. To get back to the Moon, you need money. Rich people who want to come along seem a promising source. What is more, the rich people in question are excited by the same thing as you. In a way, the space tourist is the purest enthusiast, needing no justifi-

---

[*] The Russians also charge their partners from NASA and ESA for these transport services.

cation in terms of species survival or economic resources or national pride or astrobiological insight, just the experience and memory of being and doing something in space, of seeing its sights and feeling its feelings.

And the commercialism of the enterprise offers an ideological bonus of its own. When I first started to mix with American space enthusiasts, in the early 1990s, I quickly learned that many resented the government for not having followed through on Apollo, and they were convinced that private industry must take up the challenge in order to produce a permanent presence. Some, though, went further. They argued that the purpose of private spaceflight was not merely to displace government but to undermine its raison d'être. Landing a man on the Moon had come to be seen—as Kennedy had intended it to be seen—as the ulti-mate symbol of what a government could do to achieve a stated national goal: hence the double-edged importance of "If we can put a man on the Moon . . .", stressing simultaneously what can be done and what the country chooses not to do. But what if there were no such "we"?

Private spaceflight was important to some on the libertarian right, I learned, specifically because it would remove that singular governmen-tal source of prestige, disproving the claim that there were achievements which, by their nature, were for governments alone: natural monopolies of prestige. If the private sector could land a man on the Moon—if, better still, private individuals could—government's claim to making people bigger and nobler together than they could be on their own and in freely chosen associations would be revealed as fiction.

You do not have to subscribe to such Randian nonsense to want to be a space tourist or to want to profit from the capabilities such tourism provides. You just have to have been there for Apollo and to have believed, as Jim Muncy, a long-term advocate, puts it in the documentary "Orphans of Apollo", that "because this is America, that eventually it was going to be about us." That people would be able to go to space for their own reasons.

● ○ ●

Now perhaps they will. In 2019, for the first time since Apollo, it is quite plausible that humans will walk on the Moon again within the next decade, and quite hard to believe they will not do so in the next two. There are many more people on Earth today who will walk on the Moon than who have walked on the Moon. Technology and the lure of lunar resources are part of the story of this Return, but so are politics and personalities.

The People's Republic of China has mounted a serious programme of robotic lunar exploration. It intends to follow Chang'e 4's landing in South Pole-Aitken with the first lunar sample-return mission since the 1970s. Its next stated human spaceflight goal is a permanent space station, but it has talked about following that with Moon missions. The Long March 9 rocket it is designing is the sort of Saturn-V-class booster needed for such things. As an opportunity for signalling Earthly power, the Return offers China more than it has any other country since America undertook Apollo. And it offers it cheaper. Technology is better. The risks are lower.

This in itself is enough to spur American politicians to thoughts of the Return. Though it might be intellectually defensible for a country that went to the Moon in the 1960s to treat Chinese lunar expeditions in the 2030s with a been-there-done-that insouciance, it would probably not be good politics. The desire not to seem outmatched by China is one of the reasons that, under President Donald Trump, NASA has taken on a far more explicitly Moon-focused strategy, with an overt intention to return on a permanent and eventually moneymaking basis.*

But China is not the whole story. Billionaires matter too—particularly, though not exclusively, billionaires with Silicon Valley in their background. The path down which computer technology and software have travelled since the 1970s has concentrated a great deal of wealth

---

* There are other reasons. The private-enterprise possibilities stressed by Moon advocates always appeal to Republicans. And President Barack Obama's administration was particularly Mars-oriented. On the basis that Mr Trump neuralgically shies away from any policy associated with his predecessor, a shift to the Moon under his administration would have been a sure thing even had there been no other reason for it.

into the hands of men now entering—Mark Zuckerberg, 35 in May 2019—enjoying—Jeff Bezos, 55 in January 2019—or leaving—Bill Gates, 64 in October 2019—middle age. A fair few of them still cherish the dream of spaceflight that Apollo, "Star Trek" or both lit in their hearts. Following that dream offers a way to spend money amassed from technologies closer to home on self-gratification, inspiration, ego jousting, the denting of the universe, preserving and enhancing the future of humankind, having fun, showing off and experiencing the sublime.

What more reason do you need to fly? (Tick all the boxes that apply.)

VISITS

BETWEEN 1968 AND 1972, NINE APOLLO MISSIONS TOOK 24 men there and back, with three of the men making the journey twice. Eight of the missions orbited it, six of them landed. Twelve of the men walked on the surface of the Moon, and 12 men didn't.

There have been many more visits by robot spacecraft, before and since. The first attempt was on August 17th 1958. Its wreckage is on the bottom of the Atlantic about 20km from Cape Canaveral. The following nine attempts also failed, American and Russian alike; though one American mission, Pioneer 4, did actually make it out of Earth orbit, it missed the Moon. The first successful Moon missions were Luna 2 and Luna 3, the Soviet Union's sixth and seventh attempts. One hit the Moon, as it was meant to, in late 1959. The other flew past it, also as planned. America did not manage a successful mission until Ranger 7 in 1964, six years after its first attempt.

All in all, 36 crewless spacecraft designed to study the Moon have been destroyed or sent astray by the rockets they relied on for launch and 18 have failed in space. But 58 have made it to their destination as planned.

After the Apollo missions, America sent a last robot orbiter in 1973. The Soviet Union sent a last rover in 1976. For the following 14 years, the Moon was visited only by NASA's International Sun-Earth

Explorer 3 (ISEE-3), which was not studying it but simply using its gravity to change its trajectory; the cunningly contrived path which sent it to Comet Giacobini-Zinner involved five separate encounters with the Moon.

In the decades of lunar neglect, missions flew past Jupiter and Saturn, Uranus and Neptune; asteroids and comets were seen close up for the first time. Mars, after a lull from the mid-1970s to the mid-1990s, was visited frequently, Venus and Mercury considerably less so. As these missions spread across the solar system, a picture of the Earth and the Moon in the same frame became a rite of passage for the teams controlling their cameras—a way to calibrate their equipment, to provide a startling image to the media and to renew, or deepen, the participants' sense of wonder at the endeavours they had embarked on.

The first real post-Apollo, post-Soviet Moon mission was Japanese. Despite setbacks along the way, Hiten entered orbit round the Moon in October 1991. America went back in 1994 with a military mission called Clementine which sought to use lunar science as a test for instruments developed for the Strategic Defense Initiative, or "Star Wars". NASA returned in 1997 with a small mission called Lunar Prospector. The names of both probes—Clementine, the inamorata of "My Darling Clementine", was a miner's daughter—reflected a new interest in exploitable lunar resources.

NASA has sent five further missions. The European Space Agency sent its first mission to the Moon in 2003. Japan returned with a three-spacecraft orbital mission in 2007. India launched a successful double header, an orbiter and an impactor, in 2009. It plans to launch a more ambitious lander in 2019.

The most concerted Moon programme since Apollo has been China's. Its first Moon mission, an orbiter called Chang'e 1, was launched in 2007. In 2013 Chang'e 3 landed a small rover, Yuku ("Jade Rabbit"), on the surface of Mare Imbrium. Chang'e 4 has now done the same on the farside. Of the six spacecraft operating in orbit around or near the Moon at the end of 2018, three were Chinese, three American.

Mars, at that time, boasted six working orbiters, one rover rolling across it, another rover out of contact but not yet given up for dead and a recently arrived stationary lander. Six of these nine spacecraft were American, two European and one Indian. In the decade to come, though, it is a safe bet that the Moon will once again become the solar system's most popular destination.

One source of extra Moon missions will be the private sector. As of the end of 2018, no private missions have made it to lunar orbit, let alone the surface. A company called LuxSpace, though, launched a memorial to Manfred Fuchs, the founder of its German parent company, on the booster that took China's Chang'e 5-T1 mission round the Moon in 2014.

It was the Moon's third memorial to the dead. In 1971 Dave Scott left a small memorial to fallen astronauts and cosmonauts at the Apollo 15 landing site. And in 1999 a small portion of the ashes of Gene Shoemaker were delivered to the crater that bears his name near the Moon's South Pole by Lunar Prospector, which was deliberately crashed there at the end of its mission.

Representatives of the Navajo nation objected to this desecration of the Moon, which they consider a sacred place.

- VI -

THE RETURN

THE FIRST TICKETS FOR THE RETURN HAVE BEEN BOOKED. Yusaku Maezawa, a Japanese billionaire, has purchased a trip to the Moon for sometime in 2023, though he realises that the flight may be delayed, what with the relevant spaceship not yet having been built or tested. The down payment was significant.

The trip—known as #dearMoon—is to be the simplest sort of Moon journey there is: a "free return" trajectory. The vehicle heads out to the Moon but instead of going into orbit simply swings round it and heads back to the Earth. It is the sort of trajectory that Luna 3, the Soviet probe which first saw the far side, used in 1959, and which the Apollo 13 mission resorted to after damage caused by an explosion in its service module precluded its planned orbit of the Moon.

But if the trajectory is nothing new, the rest of the trip is unprecedented. It is being paid for by a private individual. It is being purchased from a private company. It will be undertaken on a spacecraft which could, in principle, make the journey again at a later date, a number of times. It is slated to carry nine people, not three. They will not all be men. They will not all be American. And they will be artists.

Mr Maezawa—a fashion designer—has not yet said who his companions will be or what arts they will practice: musical performance and composition, poetry and dance all seem likely; maybe painting and sculpture; maybe someone whose practice is normally conceptual, or performance based. Ideally, for me, at least one outsider: someone beyond the world of the paid and the professional, the stage and the page and the gallery, one with a shunned practice all their own, demotic or hermetic. It is hard not to see something missing in a lunar art party with no unsettling touch of lunacy.

Whatever their first callings, though, all eight, and Mr Maezawa too, will have a second: acting. As Robert Lewis Shayon reflected in the *Saturday Review* after the flight of Apollo 11:

> Wherever explorers go in the future accompanied by television cameras, they will be actors, making their nebulous exits and entrances for the benefit of multi-planetary audiences. Nowhere will there ever be pure events (if ever there were); everything hereafter will be stage-managed for cosmic Nielsens.

As they look at the Moon, the Earth will look at them. Unless Mr Maezawa goes radically against the spirit of the age, there will be high-definition live feeds, social media posts, multiple angles on everything which both create and divide the common context. Unrepentant Apollo-was-a-hoaxers will troll and be trolled. On your second screen your friends will tell you what other aspect you might be watching, should be watching, have to watch again, what music you should be streaming. In 1969 Apollo's Mission Control was the first multi-screen environment that most people watching on their single screen at home had ever seen. Now all experiences can be fragmented thus.

The singular Moon will pass on beneath it all, regardless. *Aucune rancune.*

Some will be dissatisfied. The term *Media Circus* will trend. Marx's line about history repeated as farce will be dragged out. The mismatch between the "Earthrise" that surprised *Apollo 8*

> —Oh my God! Look at that
> picture over there!

and the one that #dearMoon will capture with eager well-scheduled anticipation will be much remarked on. It seems certain that some, at least, of the art will be underwhelming and that some people will delight in pointing that out.

But there should be no nostalgia for the authenticity of the past. It, too, was utterly mediated, as those lines from the *Saturday Review* remind us; just in a different way. Apollo 11 was the quintessential media event of the mass-media age, with all the attributes such events must have, many of them in space: it was planned (but not by the broadcasters), ceremonial, unique but in principle repeatable, costly, dramatic, spectacular, moving, unifying. People needed to be its witnesses to be the people they wanted to be; the media created that experience for them.

The difference with #dearMoon will be that its crew will be chosen, in part, for what the Moon can do to them and for what they might then do for those of us watching, listening, consuming, both in the moment and in the years that follow. The idea of taking artists into space is not new. After putting Yuri Gagarin into orbit on *Vostok 1*, Sergei Korolev is said to have said, "We should have sent a poet, not a pilot." The sentiment has since been echoed repeatedly. It seems ill founded to me—a confusion between poetry and the poetic. Gagarin's flight was not just a great adventure, a startling technological achievement and a propaganda coup. It was a work of art in itself—one defined more by its expansion of human possibilities than of a particular human's experiences. Would it have been improved by a deeper insight into how it had affected just one man? Eight years later, nothing became Neil Armstrong more than the modest and private life he returned to, leaving only that which he had done for the public.

For the Return, though, with the possibility already established, those objections fade. And who should lead the human Return to the Moon if not artists? The people who put in the highest bids? Space-struck volunteers chosen by lottery? Officers of the state, uniformed or otherwise? Scientists? Kardashians? Something has to happen next, and a ship of artists from around the world is no worse an idea than any other, and better than quite a few.

◐ ◯ ◑

MR MAEZAWA'S TRIP IS TO BE PROVIDED BY ELON MUSK. MR Musk has, in the past, been somewhat sniffy about space tourism. When he founded his company SpaceX in 2003 it was to do real things: to launch satellites, to sell services, to reinvent the human condition by making *Homo sapiens* a multiplanetary species. Package holidays for plutocrats were not part of the plan.

As a provider of practical services to industry and government, SpaceX has succeeded beyond almost all expectation. In the ten years since September 2008, when, at its fourth attempt, SpaceX finally launched its first satellite, the company has gone from triumph to triumph. It has more than 50 successful launches under its belt. The Falcon 9 that it was already investing almost everything in before the original, thrice-failed Falcon 1 finally flew—such faith!—now dominates the commercial-launch business. Unlike any of its competitors, it boasts fully reusable first stages that fly to the edge of space, come back down to a landing pad—or sometimes, a ship at sea—and get ready to do it again. The Falcon Heavy launcher, powered by three of those first stages yoked together, is the most powerful commercially developed booster ever; it may also be, in terms of dollars-charged-per-kilo-delivered-to-orbit, the cheapest. Since 2012 a growing squadron of crewless Falcon-9-launched space capsules, called Dragons, has been taking supplies up to the International Space Station on a regular basis. In early 2019 a reusable version of the Dragon capable of carrying seven astronauts on the same journey made its debut. In a decade the company has developed capabilities which were previously the preserve of superpowers.

This is a triumph of private enterprise; an upstart disruptor with new ambitions, new technology and new agility, developing its know-how in-house at breakneck speed, takes on the incumbents and beats them hollow. SpaceX used the best technologies it could find or imagine rather than the ones others had made do with before; it tried to continuously improve what it was doing; it brought the best of what Silicon Valley does to space technology.

But it is not a triumph of private enterprise alone. After the space shuttle was retired in 2011, the United States no longer had a way of getting cargo and astronauts to the International Space Station and back. And it had been decided that rather than develop a new government spacecraft for the purpose, the agency would encourage the development of commercial crew-transport services which it would then be able to buy. SpaceX was one beneficiary of this approach; another was Orbital Sciences, now part of Northrop Grumman, which developed a spacecraft called Cygnus for the purpose. The Cygnus, though, unlike the Dragon, is not reusable, and there will never be a crewed version of it.

This buying-services-from-industry approach—a new departure for NASA—was a tremendous success. NASA invested a bit under $400m in the Falcon 9 and Dragon, paid out in increments as SpaceX crossed 40 separate "milestones". According to an internal estimate, for NASA to have developed the rocket itself in the way it has in the past would have cost it $4bn. The same approach is now being applied to the agency's resurgent Moon programme. Small American companies such as Moon Express, Astrobotic and Marston Aerospace, with plans to take payloads to the lunar surface for paying customers, as well as for their own purposes, are competing for grants from NASA that will help them develop their lunar landers. In 2019 they will be bidding for contracts to deliver payloads to the Moon for the agency. The promise of steady cash from this anchor tenancy has already put them on a firmer footing.

If Washington wanted to procure human missions to the Moon in the same way, SpaceX could, in principle, provide surface-to-surface service in a few years with relatively little extra hardware. An architecture sketched by Robert Zubrin requires just two new vehicles. The first is a cargo carrier that you might see as being a bit like a flatbed truck. It would combine, more or less, the roles of the third stage of a Saturn V—the stage that put the command module and LM on their trajectory to the Moon—and the LM. Launched on a Falcon Heavy, this flatbed would be capable of lifting eight tonnes of cargo from low Earth orbit to the surface of the Moon.

The second new spacecraft is something like the ascent stage of the Apollo LM, but with a bigger propulsion system. Empty, this NewLM weighs two tonnes; full, it weighs about eight. Those six tonnes of propellant would give the NewLM enough delta-v to get itself up off the Moon and back into a low Earth orbit.

A bare-bones mission using Mr Zubrin's architecture—which he calls Moon Direct—starts with a crewless but fully fuelled NewLM being put on one of the flatbeds and lifted to low Earth orbit by a Falcon Heavy. A Falcon 9 then launches a Dragon with a couple of people on board. The Dragon docks with the NewLM on the flatbed, and the astronauts cross over from one spacecraft to the other. The flatbed takes the now-crewed NewLM off to the Moon; the empty Dragon stays behind in orbit.

A few days later the flatbed touches down on a grey lunar plain. The astronauts get out and do their stuff. When they are done, they blast off back to low Earth orbit in the NewLM. There they dock again with their Dragon capsule, get back into it and head home. The crowd goes wild.

A two-tonne NewLM would not provide the astronauts with much by way of comfort, equipment and provisions. But an eight-tonne habitation module, or "hab", would. If you want the mission to be more ambitious, send out an empty eight-tonne hab first on a flatbed of its own and have your astronauts land their one next to it. Send out a couple more and you have yourself a little moonbase.

Such Falcon-Heavy-enabled missions sound quite doable, quite cheap, potentially quite productive and possibly quite popular. But they fit neither Mr Musk's agenda nor NASA's.

Having developed a commercially viable system for reaching low Earth orbit, Mr Musk is on the road to Mars. SpaceX plans no further upgrades to its Falcon 9s or Dragons; the Falcon Heavy may fly only a couple of times. Instead, the company is concentrating on something which was until late 2018 known with a mix of foul-mouthed fancy and technical accuracy as the BFR.[*] The BFR is now officially called the Starship, but I suspect that many will continue to think of it as the BFR for quite some time.

Developing this new rocket is much more challenging than developing a lunar flatbed or a NewLM. It is a fully reusable two-stage system more powerful than the Saturn V, and one that needs a life support system capable of providing more astronaut-days of life in space without resupply than the International Space Station ever has.

In the end-2018 version of the design, the booster stage can lift a 100-tonne second-stage spaceship into low Earth orbit. The tanks on that spaceship will be more or less empty when it gets there. Four more such launches, using second stages kitted out as tankers, will be needed to fill up the orbiting spaceship with the methane and liquid oxygen it needs to take a fairly large crew to Mars. When it gets to Mars, the spaceship will land next to a chemical plant sent there in advance to turn Martian carbon dioxide and water into the methane and oxygen the spaceship needs to get back to Earth.

The Moon would obviously be much easier. With just a single refuelling after launch, Mr Musk's spaceship could land 100 tonnes on the Moon and still have enough propellant in its tanks to get back to Earth.

You might think that NASA, mandated as it is to go forth and exploit the Moon, increasingly happy as it is to procure services from

---

[*] In polite company, the *F* is held to still stand for "Falcon".

private industry, would be a source of judicious risk-sharing assistance to the development of this magnificent beast, as it was to the Falcon 9.

But NASA is building a big rocket of its own: the Space Launch System, or SLS. It is descended from something originally conceived of, under President George W. Bush, as a way of launching both Moon missions and the components of Mars missions. Fourteen years of development have cost, to date, something like $20bn. By the time the SLS can take to the skies routinely, if it ever does, that $20bn is expected to have grown to $30bn. It is an abomination.

When the mooncalf SLS first takes off, it will be powered by space-shuttle engines. Not engines of the same design as the space shuttle's. Used engines taken out of the three retired space-shuttle orbiters and refurbished. The thrust provided by four of those engines will be augmented by stretched versions of the strap-on boosters that were also part of the space shuttle's design. In an upgraded version, these strap-ons could be replaced by a new design using engines that are basically Saturn F-1s. The second stage will, at first, use an evolved version of the first oxygen-hydrogen engine the United States ever developed. On top of the stack will be a space capsule, Orion, of genuinely new design. That said, it is quite like an Apollo command module with more room for the crew, more capabilities in its service module and solar power—or, to put it another way, like the crewed version of the Dragon, but more expensive.

There is nothing wrong with using tried-and-true technology to get to orbit. But building a non-reusable system that will cost about $1bn a launch to do so in the age of the Falcon Heavy, let alone the BFR, is perverse. Orbit, though, is not the sole, or even the primary, purpose of the SLS. The $2bn or so spent on it every year keeps von Braun's heirs at NASA's Marshall Spaceflight Center in Huntsville, Alabama, gainfully occupied, along with a large number of private-sector contractors there and elsewhere. Senator Richard Shelby of Alabama and various of his colleagues find this congenial. When President Barack Obama's administration tried to stop NASA from building big rockets, the Senate would not let it. Though it is not strictly necessary, the senators would like the SLS eventually to have something to do; even Mr Shelby would baulk at simply lining the things up on the bluffs above the Tennessee River like Easter Island statues. So, they have assured it a role. For the time being, NASA's plans for returning people to the Moon are centred on the SLS.

If the government were to give up on the SLS, the BFR would be perfectly positioned to mop up NASA contracts for future Moon-taxi services, just as the Falcon 9 and Dragon did for space station supply runs. The fact that, since Mr Musk unveiled the BFR's design in 2016, it has been refined in such a way as to make its capabilities almost identical to those of the fully developed SLS, suggests that SpaceX is quite aware of this. But the implicit offer of something better and cheaper than what the government is building for itself is one that, for the moment, the government believes it can continue to refuse.

This is not just because of the Senate's love for the SLS. As a matter of strategy, the government does not want to be left with a monopoly provider of the launch services it deems necessary. Many in Washington—especially, I believe, in national security circles—see Mr Musk as something of a flake. It took a long time for the Defense Department to give SpaceX access to the lucrative market for launching spy satellites which is the purview of the United Launch Alliance (ULA), a joint venture of Boeing and Lockheed Martin. Just as the government bought space station resupply flights using the Cygnus as well as the Dragon, so the Dragon contract for taking astronauts to the station is paired with one for Boeing's CST-100 Starliner, an Orion-lite vessel designed to do the same job. The idea that the BFR should be America's only super-heavy-lift system is apparently quite disturbing to some policymakers, even though a super-heavy-lift capability has no obvious national security applications.

Without government support of the sort that helped SpaceX in the past, the development of the BFR, which Mr Musk sees as an outlay of about $5bn, seems too much for the company's coffers. Successful though the Falcon-based launch business is, it is unlikely to provide that big a cash flow, and its market does not have that much near-term room for growth. SpaceX has ambitious plans for a huge new constellation of communication satellites that may one day deliver many billions of dollars. But at the moment this is a competitor for investment that could otherwise go into the BFR rather than a cash cow to exploit.

Hence a new interest, on Mr Musk's part, in the Moon. When he first talked about the possibility of flying a BFR spaceship to the Moon, his tone was more dutiful, or indeed embarrassed, than enthusiastic. "It's 2017, we should have a lunar base by now," he told an audience of the faithful. "What the hell's going on?" It was vintage we-were-promised-the-Moon entitled orphan stuff. Mr Maezawa's scheme, though, offers something with more immediate charm than one day making up for the absence of the moonbase someone else should have taken care of. It offers SpaceX a way to get some development money for the BFR, an early version of which will send him and his artists on their way. Mr Musk has said, "Mr Maezawa is paying a lot of money that would help with the ship and its booster."

And he does not require the whole system. #dearMoon's free return trajectory uses so little fuel that the spaceship would require no in-orbit refuelling. Nor would it need a life support system as ambitious as that required for the long flight to Mars. It could be a preliminary, stripped-down version.

Mr Maezawa's down payment on its own is not going to pay for the BFR's development. A few more such missions, though, might make a significant contribution. Someone excited by the prospect of putting together a landing party of artists or of others might be able to do so for a billion. A simple moonbase could probably be had for less than has been spent on the SLS to date. No one should bet on #dearMoon actually taking off in 2023. But even a slower-than-promised Moon programme from SpaceX would probably bring about the Return before a programme based on the SLS, and before the Chinese, too.

AND SPACEX IS NOT THE ONLY GAME IN TOWN. JEFF BEZOS, THE founder of Amazon, one of the world's first trillion-dollar companies, is at the time of writing the richest person in the world. Since 2000 he has been regularly investing slivers of his wealth in Blue Origin, a company that builds rockets. Its first, New Shepard, is a small reusable rocket not unlike the first stage of a Falcon 9, though with only one engine it is considerably less capable. New Shepard is designed to launch a capsule in which tourists can enjoy trips above the atmosphere like those offered by Virgin Galactic.* If one or both of them launches its first passengers on July 20th 2019, I would not be remotely surprised.

New Shepard is named after Alan Shepard, whose suborbital trip in the first Mercury spacecraft, *Freedom 7*, made him the first American in space. Blue Origin's first orbital rocket, a mostly reusable system which is intended for launch in 2021, is the New Glenn; John Glenn, in *Friendship 7*, was the first American into orbit. The New Glenn will be capable of launching a bit more than a Falcon 9 and rather less than a Falcon Heavy.

After that, Blue Origin plans something in the super-heavy-lift category. The company calls it New Armstrong.

Blue Origin's accomplishments to date lag far behind those of SpaceX. It has launched nothing into orbit; it has made more or less no money. But it is clearly technically competent—the New Shepard has flown nine times apparently flawlessly—and Mr Bezos is a relentless man. He has talked of investing $1bn or so in Blue Origin every year in perpetuity. Unlike Mr Musk, he is not bewitched by Mars or unduly worried about humanity's eggs-to-basket ratio. Instead, he hopes to take part in, or indeed lead, an industrial revolution in space like the one which Gerard O'Neill and others outlined in the 1970s and 1980s.

---

* It is when such fripperies have been compared to SpaceX that Mr Musk's animus on the subject has been most clearly expressed.

Mr Bezos talks of a future a few decades hence in which a million people live in orbit, at least for some of their time, running industries that no longer have a place on Earth. In this great inversion, the economic activities that were once seen as central, the activities of making and muscle, become peripheral, and the Earth they circle is restored, through means not immediately made clear, to some mosaic of wilderness, park, golf course and garden city, a site not of production but simply of delivery.

Delivery is a business Mr Bezos knows well. And until there is stuff made or mined in space ready to deliver to Earth, Blue Origin will deliver Earthly stuff to space—not just to orbit with New Glenn but also to the surface of the Moon, using a vehicle called Blue Moon. It is designed to go into space either on a New Glenn or on some other company's rocket—one of ULA's, perhaps, or even the SLS—and land four and a half tonnes on the surface of the Moon. That is enough for it to carry a small return vehicle that could get samples back to Earth or low Earth orbit.

In late 2018 Blue Origin and a number of other concerns, including the European Space Agency and Airbus, the biggest European aerospace company, announced that they were setting up a new competition called "The Moon Race". Start-ups and small companies will compete with designs and programs to meet four challenges: to make something from lunar materials, to isolate a bottle of lunar water, to bring light to the lunar night and to grow a plant in a lunar greenhouse. The winning proposals will get money for prototyping; the best prototypes, money for development; the best developed plans, money for building hardware. The overall winners will be launched to the Moon in 2024.

Despite the name, the Moon Race is not a race to the Moon. There has already been one of those this century, and it was at best only a heavily qualified success. Announced in 2007, the Google Lunar X Prize offered $20m to the first company to land a rover on the Moon and send back high-definition video, with a variety of smaller prizes for achievements on the way to that goal. Although the original deadline was relaxed, no one won it; in early 2018, the prize was cancelled. Some of the competitors, though, are running still; SpaceIL, which is behind the Israeli mission launched in early 2019, and iSpace, the Japanese team which plans to launch an orbiter in 2020 and a lander in 2021, are Google Lunar X Prize veterans, as are MoonExpress and Astrobotic in the United States.

Instead of a race to the Moon, the Moon Race is a race to get things done on the Moon. Getting there is part of the prize, not the object of the competition. As such it marks the beginning of the commoditization of access to the Moon. By the mid-2020s, there should be a number of companies capable of taking cargo to the Moon and providing it with electricity and communication links once there. If the market looks good—if funding bodies will pay for lunar-surface science, if investors will pay for scouting out of resources, if enough rich enthusiasts just want to do stuff—then missions which bring something back will be feasible, too. From the late 2020s or early 2030s there should be a number of super-heavy-lift vehicles capable of placing ships on the Moon's surface that are able to get back to low Earth orbit, not to mention small habs. If the BFR lives up to its billing, that could happen sooner; but even if the BFR F's up, it seems a fair bet that New Armstrong or the Long March 9 will get the job done.

And once reusable rockets can reliably get robots and humans to the Moon, humans and robots will go.

Not just robots and humans paid for by governments or patrons of the arts. Private people, private robots, with private plans. The success of SpaceX, the promise of Blue Origin and the boom in innovative small satellite companies supported by the sort of venture capital that built Silicon Valley's semiconductor and software industries have created a groundswell of enthusiasm for freelance Moon exploration and indeed colonisation. Concrete plans fostered by this enthusiasm are not, as yet, fully formed or fully public. But there is an ambition for the Return which is more than just aspiration and speculation. There is a growing number of entrepreneurs with experience of doing things in space for far less than such things used to cost, along with a growing number who have billions to spend and some universe-denting yet to do. Between them they look likely to play a big role in the Moon's near future. This could be a great opening up not just of the Moon but also of space more generally.

It seems certain that more people will go into space and more companies will try to do things there. At the same time, you might worry that there could also be something of a closing down in the way that the process is being undertaken.

Elon Musk has led the most successful spacecraft development programme since Apollo; as "Chief Designer" he is said to have mastered much of SpaceX's engineering detail himself and has kept its technological developments in line with well-articulated long-term aims; he has put together a team that is able to do his very demanding bidding and has facilitated that team's consistent success. It is a truly extraordinary achievement. That Robert Downey Jr. has talked of seeing Mr Musk as a model for his performance as Tony Stark, "genius, billionaire, playboy, philanthropist" of the Marvel Cinematic Universe, might make you cringe—Mr Musk's almost endearingly stilted cameo in "Iron Man 2", partly filmed at the SpaceX factory, certainly will—but it is not all that much of a stretch. He does not have superpowers. But his technology has superempowered him to the extent that his decisions might affect history.

He is also a prick. Not an irredeemable arsehole, though he has recently looked more like one than he used to, but someone who mocks people in no place to fight back, who responds to reasonable requests with unreasonable disdain, whose self-indulgence, sometimes charming, can be gratingly self-satisfied; someone who seems willing to despise without troubling to understand, someone too easy with sycophancy and too scornful of critics. The blatant uneasiness at some public events; the maladroitness; the urge to protect things, an urge that one cannot help but trace to the damage he has talked of suffering from an abusive father: all these complicate the picture. They call forth in me, for what it is worth, a qualified affection alongside my great admiration for his achievement and my dislike of much of his recent behaviour. But for all that, and despite the fact that he has been friendly and engaged on the occasions when we have talked, he is still a prick.

I am certain that Mr Musk sees his Martian goals as altruistic. At the same time, allied with what the public knows and suspects of his character, they give people cause to worry. The idea of preserving a multiplanetary humanity from threats to the Earth is easily understood as providing a bolthole for yourself, along perhaps with your friends.

Mr Bezos (whom I have met only once—at a lunch in the 1990s we chatted briefly about science fiction stories that deal with the end of the world) is clearly more disciplined than Mr Musk. In putting his own money into buying and reviving the *Washington Post*, without as far as can be seen seeking to influence its agenda, he did a service to America and the world.[*] And in Amazon he has built a company that provides services I use on a disturbingly regular basis, all the while guiltily admiring those of my friends who, despising the company's readiness to minimise its tax burden, choose to buy things from merchants whose size and attitude they prefer.

To the extent that I can judge, Mr Bezos sees himself, like Mr Musk, as well motivated. His vision for the future might be construed as simply commercial, a plan for cosmic duopoly in which Blue Origin owns the infrastructure of the world above while Amazon delivers its bounty to the world below. But that does not seem to be his motivation. He seems to me a science fiction fan who internalized Heinlein, and O'Neill, and Greg Bear, and "Lucifer's Hammer", not to mention "The Man Who Sold the Moon": bleeding Amazon to feed Blue Origin has a touch of the D. D. Harriman to it, even if it is being done considerably more sustainably. Through his application of new technological capabilities to business, he has realised the power to make some of the things in those stories happen. He thinks that will be good for humankind as well as fulfilling to him personally.

---

[*] As well as, obviously, to journalists like me.

All that said, he has become the richest man in the world through the ruthless expansion of a company which acts in predatory and anti-competitive ways and which has long enjoyed the services of distribution-centre workers who find themselves pissing into bottles to avoid being penalized for the time taken to get to the toilet. One should not discount the possibility of prickishness.

George Bernard Shaw—who, on reading Arthur C. Clarke's essay "The Challenge of the Spaceship", applied for membership in the British Interplanetary Society at the age of 91—wrote in his "Maxims for Revolutionists" (1903) that:

> The reasonable man adapts himself to the world; the unreasonable one persists in trying to adapt the world to himself. Therefore all progress depends on the unreasonable man.

I do not think this is a universal truth. Individuals are not, in the run of things, a major source of progress in themselves, and those who are rarely become so purely to rearrange the world to their requirements. But I do think Shaw captures something about some forms of change brought about by and through companies. I once worked at a company whose founder wished it to be revolutionary. Shaw's maxim was quite prominently displayed on the desk of one of his co-founders, something to point to while rolling your eyes after a particularly unreasonable outburst. That "unreasonable man" meant, at least to some extent, "prick" was a given.

The people who get into space courtesy of billionaires need not share the foibles, the philosophies or the politics of those who open up the way. But if expansion into space is too strongly identified with, led by and reliant upon superempowered men who behave like pricks, it could lessen, sour or endanger the whole undertaking.

EXPOSURE

THE EARTH IS PROTECTED. ITS AIRS AND WATERS HOLD IN THE warmth of day and summer to comfort it in night and winter; their cooling flow drains away excessive heat. Its atmosphere is a shield, too, turning incoming bodies from outer space into the harmless splendour of shooting stars and absorbing the harshest wavelengths of the Sun's light. The magnetic field driven by currents in its molten core protects it from cosmic rays and solar wind. To travel beyond these protections to the Moon is to be exposed to the cosmos.

The exposure is not limited to the time spent travelling. Astronauts walking or driving across the cold-hearted Moon may have arrived—but they are still in space, still bombarded by swift matter and harsh energy and its extremes of temperature, still deprived of the worldly flows of a living world. The surface of the Moon bears witness to all those injuries and absences.

The constant drizzle of particle-breaking, glass-creating micrometeoroids changes the texture and colour of the regolith. It darkens, sub-millimetre crater by sub-millimetre crater, over the millions of years. Any freshly excavated material, like that which lances out in bright rays from craters like Tycho, will lose its lustre to this weathering. Tycho will be a much subtler feature on the full Moon in a billion years. But some other bright sunburst of un-weathered dust will in time take its decorative place.

The solar wind, that plasma of charged ions blown off the Sun, is another source of weathering. *Thin* hardly begins to do justice to its lack of substance; a cubic centimetre of the wind contains only about five ions, compared to 25 thousand million billion molecules in a cubic centimetre of air. It takes a few million years for a square kilometre of the Moon to pick up just a couple of grams from this wind. But thin as it is, the solar wind still packs a certain punch; its ions move even faster than the micrometeoroids do. Some of them bounce back into space, either still charged or transformed into neutral atoms by a fleeting interaction with the surface. Some stick around. They may be absorbed into the regolith as is. They may drive change.

Hydrogen ions—which make up more than 90% of the wind—may pick up oxygen. Sometimes, through this mechanism, they create water molecules. How much water is made this way, and where it ends up, has yet to be well understood. If the oxygen thus liberated comes from iron oxides, it can reduce those ores to unmodified iron. The thin layers of metallic iron thus built up make fine particles of moondust susceptible to magnetism.

The effects of the wind are made more complex by the Earth. Because the solar wind is composed of charged particles, and charged particles respond to magnetic forces, the Earth's magnetic field sweeps much of the wind away; it flows round the Earth as water in a stream flows around a rock. Some of the wind, though, is captured and trapped within the magnetic field, creating a "magnetosphere" which girdles the planet above the atmosphere.

On the Earth's sunward side this magnetosphere is compressed by the brunt of the wind; downstream, it is drawn out like a pennant in a sea breeze. Once a month, when the Moon is opposite the Sun in the Earth's sky, and thus full, it passes through this wake. Its surface, normally bathed in the positively charged solar wind, is for a day or so buffeted instead by the negatively charged plasma of the Earth's magnetotail.

As well as its wind, the Sun provides the Moon with light, including wavelengths that the Earth's surface never sees, such as very-short-wavelength, and thus "hard", ultraviolet radiation. This is powerful enough to knock electrons free from atoms, providing a static charge to the dayside as if it were amber rubbed with a cloth. The nightside is not thus charged. This explains the curious glow above the horizon recorded just before sunrise by the cameras on some of America's Surveyor landers. Charged up by the ultraviolet, the lightest dust particles repel each other and thus levitate into the sunlight, a mineral mist rising above the morning horizon.

No humans on the surface have been in a position to see the ghostly electric fingers of the lunar dawn or to have their equipment whipped by the plasmas of the passing magnetotail. The Apollo landings were all on well-lit parts of a less-than-full Moon, with the Sun well up in the sky, but not fully overhead. During the Return, though, they will surely go back for longer—for whole lunar days and whole lunar nights. They will explore more widely, looking for places which provide exceptions to the simple and established rules of the lunar regolith: don't move until you are hit; don't permit flows of energy if you can possibly help it; don't change.

- VII -

ON THE MOON

WHERE FIRST? THE OBVIOUS ANSWER IS THE POLES—THE subject of the paper on lunar-base siting by Mr Wingo that I read on the California train journey described in the introduction. But while the poles may be where the first humans go, their robot vanguard will explore more widely.

Having landed Chang'e 4 in South Pole-Aitken basin, China intends its sample-return mission, Chang'e 5, for Mons Rumker in *Oceanus Procellarum*, an intriguing plateau studded with apparently young volcanic domes that was at one point a potential Apollo landing site.

India's Vikram lander and rover, part of its Chandrayaan-2 mission, is also set to land on the nearside, at a crater called Manzinus. Only about 600km from the South Pole, it will use a ground-penetrating radar to look for buried ice. Beresheet, the lander developed by an Israeli organisation, SpaceIL, is destined for a northerly nearside crater, Berzelius, which sits between *Mare Serenitatis* and *Mare Crisium*. The site is interesting because it is a place where the crust has a faint magnetic field. These little remnant magnetic fields are an enduring lunar mystery; Beresheet, which means "In the beginning", is hoped to mark a first step towards solving it.

Beresheet's range of possible landing sites is, unsurprisingly for a modest mission, quite constrained. Were it not, it might be heading for one of the particularly intriguing magnetic anomalies that are associated with "lunar swirls", pale ribbon-like loops overlaid on the darker regolith.* The patterns and the fields seem sure to be related in some way—no swirls have been seen without magnetic fields—but as yet no one knows how. One possibility is that the magnetic fields provide partial shelter from the solar wind, and thus slow the weathering of the regolith. Another is that they are formed through interactions between the magnetic fields and dust electrified by ultraviolet radiation. Or it may be that the swirls are meteor showers in reverse.

Meteor showers in the skies of Earth occur when tiny particles of comet dust hit the edge of the atmosphere at high speed. The lunar swirls may be created when a passing atmosphere hits tiny particles of sedentary moondust just as fast. When a comet's nucleus passes very close to the lunar surface, its tenuous atmosphere (called a coma, in a comet's case) may burn up the smallest particles in the regolith.

To sort out the possibilities requires some way of getting very close to the swirls. Just landing won't work: the structure of the magnetic fields matters, and that structure is only perceptible if you move across them—the closer to the surface, the better. Hence the rather wonderfully named BOLAS mission (Bi-sat Observations of Lunar Atmosphere above Swirls) developed at Goddard Space Flight Center for NASA. It is one of many lunar missions currently being suggested that make use of very small spacecraft called cubesats.

Two almost identical satellites, Bolas-L and Bolas-H, are launched clasped together and put into a "frozen" orbit around the Moon, an elliptical path which at times brings them to an altitude of less than

---

* And not just because Arthur C. Clarke used a magnetic anomaly to signal the existence of an alien artifact in Tycho in "2001: A Space Odyssey".

15km. Once settled in, they let each other go—but remain attached by a very light tether. They line up radially with respect to the Moon, Bolas-L below, Bolas-H above. As the tether is paid out, Bolas-L begins to sink and Bolas-H to rise.

This means that neither satellite is moving at the orbital speed appropriate to its altitude. Bolas-L is moving too slowly; it should be falling towards the Moon. Bolas-H is moving too quickly; it should be flying off into space. But thanks to the tether, Bolas-H's centrifugal tendency is pulling Bolas-L up, and Bolas-L's tendency to fall is weighing Bolas-H down. The tension in the tether keeps them lined up one above the other even as they move farther and farther apart. They are like the two tidal bulges in Earth's oceans, one pulled by the Moon more strongly, one less, but lined up together.

At the end of the unspooling, the two spacecraft will be 25km apart—and Bolas-L will, at closest approach, be just two and a half kilometres above the surface. That is far lower than any stable orbit: but thanks to the tether, Bolas-L and Bolas-H are still, as far as the force of gravity is concerned, in the original "frozen" orbit; their centre of mass remains at the mid-point of the tether.

One of the things that BOLAS could study as it skims the surface is the rate at which hydrogen from the solar wind gets into the regolith and how much water-making chemistry it might get up to once there. In 2020 Moon Express intends to look at some of the same processes when it lands the first of its intended series of Lunar Scouts in a region called Rima Bode, quite close to the valley of Taurus Littrow, the landing site of Apollo 17's *Challenger*. In his fieldwork at Taurus Littrow, Harrison Schmitt identified "pyroclastic" deposits in the regolith—deposits formed when an eruption sprays lava high above the surface in something like a fire fountain, giving it time to freeze into droplets of glass before it gets back to the surface. Rima Bode, too, seems to be rich in pyroclastics, but of a different chemical make-up, and much more weathered; the ones at Taurus Littrow, which had been preserved under a lava flow then uncovered by a recent impact, were a bright and distinctive orange.

The Rima Bode pyroclastics offer potentially fascinating science, because they offer samples from deep inside the Moon. They might also be of practical interest. Paul Spudis, the American scientist who did more than any other to argue the scientific case for returning to the Moon, and sadly died before seeing it, believed that well-weathered pyroclastics with a lot of titanium in their make-up, like those of Rima Bode, were likely to be particularly good at absorbing hydrogen from the solar wind, and that the fairly uniform size of the particles would make regolith rich in them relatively easy to process. There is thus a hint of practical prospecting in the choice of the landing site. But, more importantly, there is an appreciation by the people at Moon Express of Spudis's devotion to the Moon.

Astrobotic intends to land at Lacus Mortis, which lies between *Mare Frigorum* and *Mare Serenitatis*, next to a peculiar pit that might be the opening of a lava tube. When lava below the surface keeps flowing after the surface has solidified, it can leave a long cylindrical void behind it. Most Earthly cavemaking relies on water's powers of erosion and dissolution, but lava tubes do not. They may thus be the only caves the dry Moon has to offer.

In 2015 Japanese radar studies of the Marius Hills, another set of volcanic domes in *Oceanus Procellarum*, found an intriguing double return, as though some of the radar signal was bouncing off the surface and some off a second surface some way below. NASA's Grail mission, a pair of satellites which measured the Moon's gravity field with exquisite accuracy, showed that the crust is of lower-than-average density in the same area. And there is a pit in the surface which looks as if the roof of a subterranean void has collapsed. Put it all together and you have one of the best candidate caves yet found on the Moon. And it looks big. Though Earthly lava tubes may be kilometres long, they are typically just a few metres across. The tubes in the Marius Hills, if tubes they be, could be hundreds of metres across, and maybe 75m high—twice the height of the nave of Chartres Cathedral. Computer models suggest that fast-flowing lavas and low gravity might allow some lunar lava tubes to be larger still, perhaps a kilometre or more in height and two or three times that in breadth. The low gravity is crucial; it means that the weight of the rock above the void is much less than it would be on Earth.

This is not just an opportunity for off-Earth spelunking. The tubes might be good places to live. A lunar settlement is very unlikely to be a trailer park of spacecraft-like buildings on the surface. A structure that is heated to well above 100°C and then chilled to liquid-nitrogen temperatures every month faces an alarming amount of stress and strain. And the surface is peppered not just with micrometeoroids but also with cosmic rays—high-energy protons—that the Earth is protected from by its magnetosphere. Worse, there are barrages of protons thrown off the Sun by events known as coronal mass ejections.

Over a 100-day stay on the Moon's surface, even one spent in a hab shielded against background radiation of cosmic rays, astronauts would be exposed to a 13% chance of a "solar proton event", as they are known, strong enough to raise their cancer risk significantly. There would be a 5% chance of one strong enough to cause prompt radiation sickness, and a 0.5% risk of one that would be fatal. One such particularly savage event took place in early August 1972. If the Sun had lashed out in the same way four months earlier, Charlie Duke, Ken Mattingly and John Young, the crew of Apollo 16, would have been killed. If it had done so four months later it would have killed Apollo 17's Gene Cernan, Ronald Evans and Harrison Schmitt. There are ways of providing advance warning of such events using satellites hanging between the Earth and the Sun at the Sun-Earth L1 point. But such warnings are only useful if there is somewhere to go for shelter.

A surface hab thus needs a thickly shielded inner sanctum into which the crew can retreat—something which adds to the mass. Putting your living quarters in a cave provides shielding from all such radiation throughout the living area. It also offers a pretty constant temperature, and shelter from micrometeorites, too. Hence some of the interest in lunar lava tubes. If the wormholes in the Moon's green cheese could be made airtight, they might be underground analogues for O'Neill's "Islands" in the sky, voluminous enough for towns, maybe even cities. Boa Vista, the estate of the helium-mining Cortes dynasty in Ian McDonald's novel "New Moon", is a fetching example: stretched out below *Mare Fecunditatis*, 100m across and fully pressurised, its lush vegetation watered by the spray of fountains and the streams running down its gently inclined length, studded with grand haciendas, graceful pavilions and private glades, with stairs, apartments and balconies built into the basalt walls beneath vast bas reliefs of the Orixas of the Umbanda religion and, higher still, the bright blue fusion-lit sky, it is a samba-inflected mixture of Mar-a-Lago, Rivendell, the cave dwelling of the Grand Lunar in H. G. Wells's "First Men in the Moon" (1901) and a Bond villain's lair. The parties are great.

Is Boa Vista any more unlikely in the 22nd or 23rd century than the sprawling high rises of São Paulo were 500 years ago? I cannot say (though I am pretty sure it would not be financed off helium-3 exports). For the time being, though, there is no prospect of sealing such a cavity off and filling it with air—not least because the air would freeze. Keeping a whole lava tube warm would be a power-hungry undertaking. Far easier, at least in the early years, to bring habs up from Earth, manoeuvre them into trenches dug for the purpose or modestly reshaped small craters, and then cover them with a few metres of loose regolith: they would stay a lot warmer than in a deep cave and be just as well shielded from radiation.

In time, baked-regolith bricks and melted-regolith glass might be added to the architect's repertoire. An interesting set of studies of such dwellings, built over and around inflated balloons, has been carried out by Foster + Partners; its founder, Norman Foster, is an architect particularly attuned to the simple forms of flight and space. It seems likely that most of life will still take place under such mantles, if not fully underground. The transparent surface domes which delight science fiction artists have little to recommend them. Light for the crops, which any decent-sized base will need to grow, is better provided by light-emitting diodes tuned to the most photosynthetically efficient wavelengths than by windows that see no Sun for two weeks at a time.

The likelihood that bases will be built in burrows of their own, though, is not to say that lava tubes are not worth looking into, figuratively and literally. They are cold because, like the permanently shadowed craters, they are never sunlit. That means water vapour and other volatiles released by impacts will refreeze in the caves just as they do in those craters. In general the craters are a better bet for ice miners; a very tenuous vapour will not get very far into an airless cave. But you can imagine circumstances where a particular coming together of an impact, or impacts, and a specific cave system might create something interesting and valuable, such as the once-ice-filled lava bubble under *Oceanus Procellarum* used as a scientific base in Greg Bear's "Heads" (1993).

As robotic exploration continues, expect more interest in the search for such oddities, for unforeseen structures created by chains of independent or rare events. Or, just possibly, intelligence. To expect to find alien artefacts on the Moon would be to go too far. But if there are or have ever been alien intelligences in this part of the universe, and if over the four billion years or so before humankind came along they ever visited this solar system, and if they wanted to leave some sign of their passing for some future intelligence to find, the unchanging Moon that stands close by the only living world in the system would seem the most obvious place to leave it. It was this idea, as developed in Arthur C. Clarke's short story "The Sentinel" (1951), that provided the seed for Clarke and Stanley Kubrick's "2001: A Space Odyssey". The idea served as a narrative bridge that let the film jump from the plausible near future of solar system exploration to the interstellar weirdness of higher intelligences. But both men also knew it was a good speculative idea in itself.

Do I think it is worth searching for such artefacts in a deliberate, diligent and expensive way? No. But just as I think it is worth examining radio signals from the universe at large to see if any show signs of intelligence, so I think it is at least worth keeping an open mind about the possibilities of extraterrestrial intelligence when humans and robots look for oddities on the Moon.

And oddities there will be. The processes that shape the Moon are admittedly few compared to those that shape the Earth, the raw materials they work on far more limited, the ability of the environment to push and shove and mould things in interesting ways almost non-existent. But it still has a surface bigger than Africa's, a surface that has had billions of years to develop quirks and oddities, for coincidence to pile on coincidence so as to provide truly unlikely accidents. Not all these oddities will be as easily distinguishable from orbit as the magnetic swirls or as intriguing as the lava tubes. Few if any may be of much practical use. But scientifically they may prove intriguing, at least to cognoscenti. And they might be more than that.

● ○ ●

NEVERTHELESS, AS THAT PAPER BY MR WINGO SHOWS, THE poles—the Moon's most striking oddities—remain the best bet for human bases early on in the Return, and perhaps even for its first human landings. This is for reasons of practicality, potential and politics.

The practical reason is power. Power on the Moon is going to be solar or nuclear, and as yet there are no suitable nuclear options. Lunar nuclear reactors would need to be very light, by the standards of such things, if they were to be shipped up from Earth. They would need to work with little if any water (of which most power reactors use quite a lot) and to have very low maintenance requirements. They would also have to be so safe that a government would be willing to license one being launched from its territory.

For the moment that leaves solar. And for solar power, as for crops, 14-day nights are a problem. A solar-powered lunar settlement would need enough panels to provide more than twice the power it needs during the day and enough batteries to store the unused half of that power for use in the night. That's a lot of capital expenditure.

After working on a study of the costs of solar power on the Moon, Geoff Landis, a NASA engineer who is also a poet and science fiction writer, took to fiction to explore a quirky alternative to batteries:

mobility. "A Walk in the Sun" (1992), one of the nicest lunar itera-
tions of science fiction's perennial resourceful-individual-against-the-
hard-facts-of-the-cosmos trope, tells the story of Trish Milligan who,
having survived a crash landing on the Moon, needs to hold out for
a month as a rescue mission is mounted. She has a spacesuit with big
solar panels and a lot of protein bars, but not much by way of batteries.
So she decides there is nothing for it other than to walk all the way
round the Moon, keeping pace with the Sun.

Unluckily for her, she crashed near the equator, meaning she has to
walk a distance the same as that from New York to Los Angeles at an
average of 16 kilometres per hour if she is to stay ahead of the night-
edge. Easy enough when bounding along at a sixth of her terrestrial
weight on the smooth maria; harder in mountains and the unrelenting
highlands of the farside.

If she had crashed at a higher latitude, she could have set an easier
pace. Up at 70°N, around the latitude of *Mare Frigorum*, so named
because it is the Moon's northernmost maria, you can stroll west at
less than six kilometres an hour and keep the Sun on your shoulders
forever. And right up at the poles there are places where you hardly
need to move at all.

The Moon's upright posture with respect to the plane of the ecliptic means that, at its poles, the Sun sits near permanently on the horizon. This tangential lighting is what allows the depths of polar craters
to be forever cold and dark; it also allows polar uplands near perpetual
day. These sunlit uplands are not a new idea. In "The Moon" (1837),
Wilhelm Beer and Johan Madler pointed out that the Moon's obliquity meant there could be places at the poles that saw little or no night;
Claude Flammarion, taking up the idea some decades later, dubbed
such places "peaks of eternal light". Robert Goddard wrote about the
possibility, too, as well as that of permanent cold spots in craters below
the peaks. Now both types of feature have been seen and quantified.

Today's maps of the Moon show that there are raised areas of,
if not eternal, then very long-lasting, light at both poles—ribbons
of raised land on crater rims which see the Sun more than 80% of

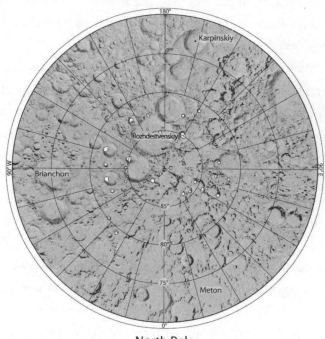

North Pole

South Pole

  Permanently Shaded Regions

the time, and where solar panels mounted vertically, like the sails of a ship, could realistically provide three times the average power of systems elsewhere that would get sunshine only half the time. For a solar-powered moonbase, you would need a very good reason not to go to one of the poles.

Beyond power, the poles also offer potential, in the form of those frozen volatiles. The presence of water in significant amounts will, other things being equal, make it easier to sustain a moonbase and to refuel rockets there. There are other sources of propellant on the Moon—you could pair solar-wind hydrogen absorbed in the regolith with oxygen torn out of various minerals. But though hydrogen is thousands of times more concentrated in the regolith than helium-3, you would still need to process quite a lot of regolith for serious amounts. And pulling apart minerals for their oxygen takes a lot of energy. Energy itself is cheap on the Moon, at least during daylight; the capital equipment needed to gather it up and use it will probably not be. Hence the attraction of the volatiles at the poles that can be liberated with the equivalent of a kettle.

The case for landing near the ice is reinforced by the fact that if you have a single moonbase and a supply of fuel, all the rest of the Moon becomes easily accessible. As Robert Zubrin points out, the delta-v needed to get a NewLM lunar lander back from the Moon to low Earth orbit is also enough to have it take off from one place on the Moon, land at more or less any other place, take off a second time, and land again back where it started. You can repeat the procedure for some other destination as soon as you can scrounge together another six tonnes of propellant.

This is not the way that NASA is currently planning to do things. When, under President Obama, the space agency developed the idea of capturing a small piece of asteroid and studying it in space—an idea sold, in an unconvincing way, as preparation for a trip to Mars, and seen by many as an excuse for not committing to a Moon mission—it imagined building a small space station near the Moon to serve as the asteroid wranglers' base. When President Trump's pivot to the Moon saw the asteroid plan dumped, this Deep Space Gateway became the Lunar Orbiting Platform-Gateway, and its purpose changed to lunar exploration. Now it is just the Gateway. But it survives, as do most NASA programmes on which money has started to be spent. And it is still a bad idea.

The Gateway is a much smaller version of the International Space Station that is much farther away. The same partners are involved, at least so far. It is American led, but the Europeans have said they will provide some hardware, as have the Canadian and Japanese space agencies; whether the Russians will stay the course is hard to say. The power system is to be bought from the communications-satellite industry. The first habitable components are, in theory, to be launched in the mid-2020s by the SLS.

Thus, in assembling a space station from prefabricated components, the Gateway programme recapitulates something done more ambitiously two decades before, but this time on a smaller scale in a much less accessible orbit as a way of justifying the use of an otherwise pointless rocket which will probably not be ready on time.

Staging missions to the surface of the Moon from the Gateway makes them more complex than they would otherwise need to be but provides no clear benefit. Before the Gateway lost its asteroid rationale, few if any experts thought that a stopping-off point somewhere between low Earth orbit and the lunar surface was a useful addition to plans for lunar missions. The Gateway crew on board could, admittedly, control tele-operated rovers on the lunar surface with less of a time lag than if they were back on Earth. But as they twiddled their joysticks they would be soaking up a lot of cosmic radiation from which they could be protected by the nice regolith walls of a moonbase.

This is not to say that a spacefaring future will have no need of orbital infrastructure. If people, or their robots, are to do more things in Earth orbit—assemble large new satellites, repair ailing old ones, manufacture weird materials in microgravity, enjoy honeymoons with out-of-this-world views and microgravity sex, build space weapons that can shoot down other people's satellites, disassemble small asteroids, whatever—they will benefit greatly from a set of depots where spacecraft in need of delta-v can pick up propellants. Going back down to Earth to refuel means having to spend about nine kilometres a second of delta-v just to get back to orbit. It also means that your spacecraft needs to be tough enough to withstand the stresses of re-entry and launch from Earth. Spacecraft that stay in space require less mass for their construction, and thus less propellant for a given amount of delta-v. Like the LM, they can be delightfully flimsy.

Propellant depots in orbit are, at the moment, the main export market imagined by the sort of companies and enthusiasts who talk about making a business out of lunar resources. The rationale is that it is a lot easier, in delta-v terms, to get to low Earth orbit from the surface of the Moon than it is from the surface of the Earth. One study by ULA estimates that fuel delivered to such a depot might be worth $3,000 a kilogram. A moonbase which exported 300 tonnes a year would thus be turning over the best part of a billion dollars. If you could build a bargain basement moonbase for $10bn or so, it might, if you squint a bit, pay little attention to the risks and cross your fingers, look like a potentially profitable proposition.

If you can do it at scale, you do not even need to burn any of the precious lunar fuel in order to export it. Imagine a factory on the Moon making little barrels with heat shields and tiny engines, filling them with propellant and throwing them towards the Earth with a mass driver like the ones that Gerard O'Neill popularised in the 1970s. The mass driver, being solar-powered, needs no fuel at all. The heat shield is used for aerobraking—skimming into and out of the Earth's upper atmosphere so as to lose velocity without being subjected to the rigours of full-on re-entry. The barrel having thus slowed down, the little engine nudges it into low Earth orbit, where a fuel depot empties it of its burden. Its metal structure is sent back to the Moon to be reused or recycled.

Alternatively, create an orbiting tether like the one in the BOLAS mission, but longer and in a higher orbit. Then start it rotating around the centre of mass. With the right rate of rotation and the right length of tether in the right orbit, you can pick things up from just above the surface of the Moon when the tether is vertical and throw them towards the Earth as it swings up and around. If the throwing arm catches the returns, the system's energy needs might be minimal.

This all means it is possible to export fuel from the Moon for quite a low marginal cost. But to do so in bulk, you need to invest a fair amount of capital—not just to mine the ice and volatiles and refine them but also to build your mass driver and its power plants, or put together your spinning tether system. And that is why the Earth, for all its delta-v handicap, looks likely to keep the Moon out of the fuel business. Both liquid oxygen and liquid methane are pretty cheap on Earth. The fact that you have to use a lot of them to lift a bit more of them into orbit is not the problem it might seem.

It has been suggested that, once the development has been done, the unit cost of a BFR might be somewhere in the region of $300m to $400m, which is what a big modern airliner costs. If you got 100 flights out of it, you would have a cost per launch of about $7m. If you chose to lift 150 tonnes of fuel to a low-Earth-orbit depot with that launch, you would be spending about $50 a kilo to get it there.

It is possible to imagine a moonbase might be a competitive source of low-Earth-orbit fuel at $3,000 a kilo. But at $50 a kilo you need an imagination that extends to unicorns and fairy dust. Obviously, the same might be said of a BFR that performs so miraculously well. But a system 50 times costlier would still undercut that price point quoted for fuel from the Moon.

Hence the fundamental paradox of commercial moonbase development. It is not really possible to build one without the low-cost ways to reach Earth orbit that systems like the BFR seek to offer. But if costs get that low, then the Earth will own the market for low-Earth-orbit fuel. If costs stay high, then maybe a moonbase can turn a profit on selling fuel. But if getting to orbit from Earth stays expensive, then people will not do it very much, and the market for on-orbit fuel will remain small.

If, like Gerard O'Neill, you can imagine a use for tens of thousands of tonnes of very cheap lunar materials in another orbit, then mining the Moon may make sense. If you can find something on the Moon that is economically transformative, the same might be true. But if you want the Moon to provide moderately high-margin goods which are readily available for launch from Earth, you will be disappointed.

The uses of space that have so far made money have all been Earth-centric—communications, remote sensing, navigation services. To the extent that space can support the industrial dreams of someone like Mr Bezos, this has to continue to be the case. That means that, contrary to the stance of the true believers, space will not be an alternative to the Earth, a step beyond it. It will be an extension of the Earth. And in that world, there may be little room for an economically relevant moonbase.

● ○ ●

THE FACT THAT THEY MAY NOT BE APPEALING AS A SOURCE OF export earnings does not mean that the volatiles at the poles are not interesting. A moonbase with local access to water and other volatiles is a much sounder proposition than one which needs to be supplied with everything from Earth. That is why politics, too, argues for putting your first base at one of the poles, and possibly hastening to make the first crewed landing of the Return there. To do so would be a way to stake a claim.

The states party to the UN Outer Space Treaty, signed in 1967, agreed to keep space from becoming a place of national rivalry, recognising it instead as "a common province of mankind". The treaty put no restrictions on going into space, as long as it was for peaceful purposes. The states retained sovereignty over any spaceships and extraterrestrial bases they might use to further those purposes, in the same way that their national law applied on their ships at sea or their bases on (similarly stateless) Antarctica. They also had to take responsibility for any spacecraft or bases owned by their citizens or launched from their territory. But they themselves had no right to claim sovereignty over any celestial body, or any part of one, whether by declaration, by occupation or by other means.

That meant the states could not claim mineral rights anywhere beyond the Earth. But what of non-states? Typically, the rights to exploit mineral resources are governed by national law, and without sovereign claims such law would not be settled. The problem was left to future treaties, conventions and protocols to sort out.

The Moon Treaty of 1978 was the first attempt to do so. When it came to resources, the Moon Treaty made the Moon "the common heritage of all mankind", a status which precluded its exploitation for purely private gain. Any economic benefits flowing from the Moon had to be shared. The treaty recognised that the nations responsible for letting those benefits flow would have a special claim on them, but so would developing nations which had no way of competing for such resources. To redress, over the long haul, that lack of wherewithal, the treaty included the idea that the technologies which permitted such development should be shared, too. "The Moon Belongs to Everyone", as the song has it.*

President Jimmy Carter's administration wanted to ratify the Moon Treaty, and thus become bound by it. Proponents of private enterprise, in America and elsewhere, objected to it. One of the first campaigns the L5 Society undertook was to block the treaty's ratification. Whether or not its efforts made a difference, the treaty did not get ratified. The Soviet Union refused to play ball, too. When the Moon Treaty struggled into force in 1984, its strictures bound only its 18 ratifying states, none of which had an independent space-launch capacity.

The terms of the Moon Treaty would not make commercial exploitation of lunar resources impossible. The UN Convention on the Law of the Sea, signed in 1982, applies the same concept of "the

---

* In "The Man Who Sold the Moon", Harriman gets that song back into the charts through a payola scheme in order to smooth the way for the UN to claim sovereignty over the Moon, having arranged in advance for it to license the actions and claims of his own Moon company.

common heritage of all mankind" to resources on and below those parts of the seabed beyond continental shelves and territorial waters. This is one of the reasons why the United States Senate, despite being urged to do so by presidents of both parties, has never ratified the Convention on the Law of the Sea, either. But the convention's application of the common-heritage principle does not preclude commercial development.

The Convention on the Law of the Sea created the International Seabed Authority as a way to spread the benefits of mining the ocean floor while seeking to maintain incentives for investment. The convention's critics draw a direct line between the creation of this authority and the fact that, almost half a century after people first began to enthuse about ocean-floor resources such as manganese nodules, there are still none of them on the market; they blame rules, bureaucracy and parasitic expropriation. They may have a point. But an alternative, or at least complementary, view is that the technological innovation required was hard and the demand for sea-floor minerals not all that great. Today a number of companies are sending robots to the floor of the Clipperton Fracture Zone in the eastern Pacific to explore the possibilities of harvesting manganese nodules. The International Seabed Authority approved those efforts, is monitoring their environmental impact and will reap some of the eventual benefits, if any. But so will the companies. That is why they are doing it.

It might seem pretty fruitless for the states party to the Moon Treaty to set up a similar regime, given that none of the nations with the capability to get to the Moon are among them. And such a regime might stymie resource exploitation if it was poorly thought through, or if that was its intention. In other circumstances, though, a legal regime might encourage such developments. People making big investments normally like to do so on a firm legal footing, and an International Lunar Authority might provide such a thing. It could also serve what may become the important role of arbitrating between different interests.

Some examples. There are radio astronomers who want to build instruments on the lunar farside, the only place within light years that is permanently shielded from the Earth's racous radio emissions. But other scientists on the farside, or for that matter ice miners in South Pole-Aitken basin, might rather like to use radios of their own. Who decides what they get to do?

Solar physicists have realised that just as today's regolith preserves today's solar wind, so past regoliths preserve solar winds of the past—which might be intriguingly different. Sampling parts of the maria where there is a stratigraphic sequence—where the top of a lava flow was pummelled into solar-wind-absorbing regolith, absorbed some solar wind and was then covered over by a second flow of lava, which produced a second layer of regolith, and so on—might provide a record of hundreds of millions of years of solar activity. But this laminated regolith might also store hydrogen and other goodies particularly well. If such sites are rare, do they get preserved for the scientists or exploited by the miners?

Until quite recently, such never-never problems only worried obsessives. But now the Return looks reasonably imminent. And the mapping of resources from orbit which has made the Moon look more promising in terms of resources has also made it look much smaller.

The Moon's total area may be greater than Africa's, but even generous estimates of the permanently shadowed regions at the poles give those of the North Pole, added together, an area just a little bigger than that of Gambia, the smallest nation on that continent; those of the South Pole are roughly the area of eSwatini, formerly Swaziland, the second smallest. Put another way, the combined cold traps of both poles cover slightly more ground than the Greater Houston metropolitan area. And if the six or so Peaks of Eternal Light that offer sunshine for more than 80% of the year cover an area greater than that of Houston's six largest shopping malls, I'd be surprised.

The areas around the peaks and close to the shadows are thus in very short supply. They are also quite possibly subject to conflicting interests. The slowly accumulating volatiles there might be treated as an economic resource; they might also be treated as another valuable stratigraphic record, one recording impacts, and perhaps other processes, over the whole history of the solar system. And mining at one site might, if it produced a lot of stray vapour, contaminate others.

These are not necessarily big problems. eSwatini is only a tiny sliver of a country, but it is still quite a large place. More than a million people live there. Solar panels on Houston's Galleria mall alone could provide tens of megawatts. But small problems are still problems, and when there is no way to settle them, they can grow. The Outer Space Treaty, by which every country which might get to the Moon is bound, gives science precedence over other activities. But it provides no mechanism for solving disputes over what is real science and what is an attempt to cover the land with something that has a specious scientific rationale simply in order to lay claim to it. Nor does it have a way of offering either relief or intervention.

In the past, it might have been possible to come to some sort of international agreement that would set up a framework for resolving disputes and allocating some sort of rights, whether under common heritage or some other principle, building on the Moon Treaty. It might just be a matter of registering limited claims to specific areas for specific times and abiding by rules about contaminating other places. It might be something more profound, such as ruling one pole to be the place for exploiting the ice and the other the place for studying it. Either way, though, the world is less capable of such things than it was. International institutions are on the defensive, not stretching their capabilities. The United States, which invests more in space than any other country, has more capabilities in space than any other country and has more space-focused entrepreneurs and start-ups than any other country, has long had a distrust of treaties, in part due to the fact that their government can be sued by its citizens if the treaties are breached. The idea that American entrepreneurs who make it to the Moon should be able to exploit it for their own profit is not a remotely controversial one in Washington. Many Americans would see it as axiomatic.

But that does not necessarily mean that the Moon Treaty has no power. If China acceded to it, for example, and led a round of negotiations aimed at establishing a new lunar authority—an authority it would surely seek to dominate—it would become a much bigger issue. If India, France or Japan joined in the process, it would be bigger still. Jan Woerner, the director general of the European Space Agency, champions what he calls the "Moon village", a vision of relatively small-scale settlement in which different public and private ventures co-operate, either on a shared site or in a more distributed way, in the development of a lasting human presence on the Moon. Common goals, common values, common technical standards: it is an appealing vision. It would surely be even more so in the presence of a clear, permissive legal structure.

In the presence of such a structure, control of the facts on the ground would still matter—but it would do so all the more in the absence of such frameworks. That is an argument for landing early crewed missions, perhaps even the first crewed missions of the Return, at one of the most appealing polar sites and for building some sort of base there, rather than sitting in an orbital Gateway plopping down landers for short stays now and then. And it is as much an argument for private corporations to do so as it is for the Chinese or American governments. Such a rush does not preclude an international agreement at a later stage—the Antarctic Treaty recognises, in its way, the claims of states that had bases on the continent before it came into force. It also allows for commercial development of the continent's natural resources, if the parties all agree, in 2048, on the terms under which that could go ahead.

But the possibility of a rush for property does provide a further argument for not being a prick, whether at a personal, corporate or national level. Scrappy entrepreneurs giving the finger to the UN to get rich Moon mining will sound good to many. A multi-billionaire getting a stranglehold on a planet's worth of resources that should be a common heritage for all humankind plays less well.

Perhaps it is of little significance. Lunar resources are a big issue if you want to make a moonbase permanent. At the scale of national economies they hardly move the dial. Disputes over who has the right to go where in the South China Sea could escalate into a real concern very quickly; rights of passage across Shackleton Crater seem less likely to.

But in this respect the Moon does not exist in a vacuum, responding only to the solar wind. The political climate on Earth—and thus, in the Anthropocene, the Earth's own climate—will buffet it, shake it, reshape it. The way people treat each other in Shackleton Crater will depend on what happens in the South China Sea. The extent of any future development of the Moon will depend in part on what is found there, in part on ingenuity, in part on luck. But it will depend much more on how the rest of the world develops, both politically and economically; what its wants are and what its conflicts. Again, space is an extension of the Earth, not an exemption from its strictures. An antagonistic world will create a Moon to match.

◐ ○ ◐

A VITAL PART OF WHAT THE ORPHANS OF APOLLO, AND YOUNGER space enthusiasts, want from the Return is that this time it should be permanent. If there is a line they quote even more than the one about eggs and baskets, or the one about not because it is easy but because it is hard, it is one from Konstantin Tsiolkovsky himself: "The earth is the cradle of the mind; but one cannot stay in the cradle forever." Expansion into space is a transition from which one should no more turn back than the transition to toddlerhood.

Hence the interest in whether law stifles or enables commerce and investment, and what sort of legal environment, if any, limits conflict while allowing growth; those are the sorts of things that might make settlement sustainable. Hence the driving interest in making space pay. If people profit from space, they will stay in space. If the world depends on there being people in space, it will arrange to keep people in space. If Moonpeople can sell rocket fuel to satellite operators, or rock to L5-ers or helium-3 to fusioneers, then they can stay up there strip mining the regolith and rooting out the polar ice to their hearts' content while serving the needs of human destiny.

Leaving aside, for the time being, whether those are reasonable business propositions, the idea of extractive industry as a route to sustainability may sound historically ill-informed, or indeed self-servingly deceitful. But in space, say O'Neillians like Mr Bezos, the Earthly contradiction between exploitation and sustainability need not apply. When you are through digging up the four billion years of history at the Moon's poles, why not move on to one of the icy moons of Jupiter? It's all just a matter of delta-v. Space cannot be depleted of its resources. The frontier is endless.

There is more than one problem with this, though. Humans are not just minds. They are bodies. This means that they require carbon, nitrogen, phosphorous and dozens of other elements if they are to eat and stay healthy. All these elements can be found in space. But some, including carbon and nitrogen, are in very short supply on the Moon. This means either that recycling on a moonbase has to be very good indeed (which will be challenging, especially in a small base) or that such bases remain continuously supplied from elsewhere.

It also means humans need protection from their environment, as well as by their environment. Take moondust. The fine grains of the regolith are largely composed of shards of glass a few microns across—not quite the last stuff you would want in your lungs on a regular basis, but not far off. It damages mechanisms, too. By the time Harrison Schmitt and Gene Cernan got back to the *Challenger* after their third sortie onto the surface, the seals between their suits and their gloves were sufficiently damaged by the dust that a fourth might have been too dangerous. And the dust will get into the base on clothes, on vehicle tracks and whatever. Systems to deal with this through air baths, multiple airlocks and the cunning use of the dust's magnetic properties should allow a multi-layered defence-in-depth capable of keeping the living areas of a hab comparatively dust free. But keeping them that way may require eternal vigilance, and as in most pursuits with such requirements, standards may lapse over time.

Perhaps most profoundly, human bodies have been shaped by evolution to and in an Earthly environment. Take them out of it and

they change. Experience on the International Space Station has shown
how, in microgravity, fluids redistribute themselves through the body,
with pressure in the skull flattening the eyeball and compressing the
optic nerve. Blood volume is reduced; the heart weakens, as do other
muscles; bones demineralize. Most of this seems reversible, and some
of it can be forestalled through prophylactic exercise and possibly with
drugs. But it is not a minor thing.

No one knows what long-term effects lunar gravity has. No Apollo
astronaut stayed on the surface longer than three days and three hours.
No non-human animal has spent any time on the Moon at all. One
might assume that the Moon's effects are less severe than the effects of
microgravity, maybe much less severe. But no one knows.

Science fiction has long speculated that the effects of living on the
Moon might be damaging for those who start off well, but good for
those who are ill. A body in which the heart has less work to do and
the muscles and skeleton have less weight to support will weaken from
lack of challenge; a body not up to the challenges of its weight or of
pumping its blood might get a new lease on life under the same circumstances. Thus the Moon, or a low-gravity space station, might be
a bad place for fit young astronauts but an ideal place for people who
are elderly or infirm.

This seems more than a little optimistic—why should the dep-
redations of low gravity not diminish the capacities of the unwell as
much as they do those of the healthy, and to their greater detriment?
But I suppose it is a possibility. It is also conceivable, as Clarke speculated in "The Secret" (1963), that the healthy could benefit from
having bodies overspecified for the environment they find themselves
in, and that life-spans might be greatly increased on the Moon. That
would clearly be a transformative discovery, especially if it led to an
exodus of the wealthy, thus reinforcing their prickishness in the minds
of the left-behind. Again, though, it seems unlikely. That cardiovas-
cular systems might stay intact for longer with less to do is perhaps
plausible. But why should cancer, or dementia, be lesser threats to
lighter people?

Going in the other direction, people used to the Moon tend not, in science fiction, to do well on the Earth. They are oppressed by six times the weight that they usually carry. They are assailed by pollutants and diseases that their carefully managed environments have excluded. The cost the spindly emissaries from the sky pay for their return is portrayed as a sacrifice, sometimes the ultimate one. Ben Bova's "Millennium" becomes a passion play in which the exoskeleton Chet Kinsman wears as he returns to Earth is effectively the cross on which he dies.

And what of the nativity? In 1962 Clarke published a short story under a title which mixed Tsiolkovsky with Walt Whitman. It ends with the narrator, in charge of building spaceships on the Moon, listening to the radio:

And then, over all the Moon and half the Earth, came the noise I promised to tell you about—the most awe inspiring sound I have heard in my life. It was the thin cry of a newborn baby, the first child in all the history of mankind to be brought forth on another world than Earth.

We looked at each other in the suddenly silenced blockhouse and then at the ships we were building out there on the blazing lunar plain. They had seemed so important a few minutes ago. They still were. But not as important as what had happened over in Medical Centre, and would happen again billions of times on countless worlds down all the ages to come.

For that was the moment, gentlemen, when I knew that Man had really conquered space.

Leaving aside the "Really? '*Man*'?" that comes immediately to the 21st-century mind, Clarke's "Out of the Cradle, Endlessly Orbiting" is one of those bits of science fiction which hits on a question of real and practical importance. Whether human pregnancies can come to healthy term on the Moon—or, for that matter, Mars—is a fundamental question. It is also one that space agencies, Moon enthusiasts, science fiction writers and billionaire visionaries have treated with little more than lip service, and frequently not even with that. Clarke's use of "Man" points to something fundamentally gendered about a search for the future of the species which depends on BFRs plunging into space. Satirising this in his story "The Big Space Fuck" (1972), Kurt Vonnegut imagined a rocket devoted entirely to the launching of freeze-dried human sperm. It was called the Arthur C. Clarke.

All people ever born have grown, first in the womb then out of it, in Earth gravity. Evolution has undoubtedly built that amount of gravity into processes, such as the way that the head engages with the cervix as birth begins. It may not be a vital piece of the process of growing a baby. The unfolding of pregnancy can adapt itself on the fly to a wide range of conditions; the same programme, after all, works in women of many sizes and shapes, and a wide range of babies too, including those which come two or three at a time. But there is no reason for evolution to have built in a flexibility that extends to the conditions on the Moon. If low gravity, like microgravity, leads to changes in internal pressure and the redistribution of fluids throughout the body, why should the amniotic fluid be an exception? What difference might changed stresses and strains make to growing bones and muscle, either in the womb or in the months and years that follow. Such effects may be much less pronounced in a sixth of the Earth's gravity than in microgravity. Still, will they really mean nothing to a fetus, or to a growing toddler leaving the non-metaphorical cradle?

Walter Miller's "The Linesman" (1957) is one of the earliest tales of a blue-collar Moon where infrastructure is built by the same sort of men who laid the railway tracks and built the bridges of Earth. The routine of the men stringing power lines and tramways from Crater City in Copernicus to an isolated mine is disrupted by the arrival of a spaceship full of prostitutes responding to the market opportunity provided by the Moon's lack of women. They are absent because, as Novotny, the foreman, reminds one of the horny men on his work crew:

You can't raise kids in low gravity. There are five graves back in Crater City to prove it. Kid's graves. Six feet long. They grow themselves to death.

Biology is not destiny. If women cannot bear children in low gravity, or if children cannot grow up healthily in such conditions, perhaps confinement and childrearing will in the end take place in orbiting space stations spun up to provide enough gravity—or back on Earth. Perhaps there can be a stable Moon settlement even if pregnancy, labour and childhood have no role there. Perhaps biological science could, with the sacrifice of enough experimental primates on the Moon, solve the problem, if problem there is.

But it may well be a problem, and so far it has been one almost universally ignored in speculations of a lunar future. Whatever the technologies of the Return, whatever the priorities which press for it and the laws which govern it, whatever the property rights which permit it, it may yet be the case that no one will ever be a natural-born Moonperson.

STORIES

THE STORIES THEY TELL OF IT IN DIFFERENT PLACES HAVE LITTLE IF anything in common; they frequently go against one another. Indeed, even in one place, or among a single people, they may vary, or even contradict each other. Its power may be feminine, but its face the image of a man; it may both impregnate and give birth, tell lies and tell truth. Some of its stories may be the sort of stories that you tell to strangers just to see what foolishness they are gullible enough to believe—we scare it away with sticks during the eclipses, lest it steal our animal skins! Other stories of the Moon may be of the secret sort which can never be told to anyone who does not already know them—which sounds impossible to those who know no such stories, but is something at which others just nod, and keep silent.

It is believed by some—and here is a story about stories—that the first tales of sky gods were tales of the Moon. The Moon, after all, is a character, with a storyline; it counts the days; it is a friend to the hunter and the raider, providing light to see by and dark when the dark of the Moon is needed. The Sun is more powerful, and its seasons matter more to the farmer. But its power is less that of a person. And the farmer came later.

Some stories of the Moon's influence are baked into language—in words like *mania*, and *lunacy*, and *menstruation*. This does not make those stories true. Women's monthly cycles are close to the duration of the Moon's, but they are not synchronised to it. Nor is madness, at least as measured today, which shows no relation to the phase of the Moon. That said, is there anything madder than a light that makes things look different and only comes in darkness? And if there is a cycle of the womb and a cycle of the sky, do they have to be in lock-step to be some sort of same? Stories can be true and untrue at the same time; they can mean one thing and also its opposite. Think of the little story engraved on a metal plaque in the Sea of Tranquility: "We came in peace for all mankind."

There is nothing constant in Moon stories, except perhaps change. For example, despite its cycling, the Moon is not, as some believe, a universal symbol of the womanly or feminine. In some places the Moon is male, in others female—it is not obvious that either gender predominates. And, yes, the Moon is often canine, or lupine. But it can be feline, too. Freyja, the Norse Moon goddess, has her chariot pulled by cats.

It is generally true that, when the Moon is male, the Sun is female, and vice versa—"Brother Sun, Sister Moon", as the Franciscans have it. Often the two are siblings. That said, they are also often lovers, and not infrequently both. But even this binary is not universal. For the Tukano of the upper Amazon they are both men, one of whom has taken the headdress from the other. In Dahomey, Mahu and Lisa, the married children of Nana Buluku, who made all things, share jurisdiction over both Sun and Moon.

The incest of Sun and Moon, when it occurs, may or may not be consensual. In a story from Greenland, Anningan, the Moon, assaults his sister Malina, the Sun, in the dark of her chambers. Malina, to find out who is attacking her, rubs soot on his unseen face without him noticing. That is why the Moon is mottled, and how Malina learned of her brother's guilt. There are other accounts of this matter, though, as of all matters.

There always needs to be some story for the cycling of the Moon. Anningan regularly wastes away until eventually he must leave the sky to hunt for seals. On his return he fattens himself up. The Bantu, or at least some of them, say that the Moon's thinning and fattening is down to his two wives, who are the evening and morning stars. One of them—I know not which—is a poor wife; when the Moon is with her, she does not feed him and he fades away. The other is a good wife; when he visits her, she feeds him back up to full again. It is said that those who tell this story do not know that the morning and evening star are the same body. I am not so sure. I think two people behaving according to their individual natures can easily be one person behaving according to more than one of hers. I would be surprised if others do not think this too.

A sadder story of the Moon's departure and return is told by the Masai. Le-eyo, the first man, was taught that at the time of someone's death, he must dispose of the body while saying, "Man, die, and come back again; Moon, die, and remain away" so that the dead can be reborn. But when the first death came, that of his neighbour's child, he misspoke, saying, "Man, die and remain away; Moon, die and come back." And that set the way of things for all dead children, including Le-eyo's. The Moon is reborn, month after month—but the dead have, ever since, stayed dead.

Sometimes married to other planets, the Moon may also be allied with them more subtly—particularly with Venus and Mercury, planets which often leave the sky, as the Moon does. Moon goddesses renowned for beauty are quite often goddesses of Venus, too; Ishtar of Babylon, with her crescent headdress, is one such. So is Frejya of the cat-pulled wagon; her day is Friday or Freitag in northern Europe, Vendredi, Venerdi and Viernes—days of Venus—in the south. But it is the same day, regardless. Trickster Moon gods, of whom there are many, are for their part often linked to Mercury. Like him, they have a habit of appearing at the crossroads—moonlit and multifarious, places of choice and divergence

Kalfu, a *petro* of Papa Legba, who controls all crossings between this world and the world of the spirits, is such a one. Papa Legba—who, back in West Africa, before he came to Haiti and the rites of Voodoo, was most definitely a trickster—is wise and can be kindly, though you would be ill advised to trust him. Kalfu, also known as Carrefour, is, like Voodoo's other *petros*, demonic. Not the pale Moon; the dark Moon, red and spiteful, drunk on a mixture of rum and gunpowder. He lets chaos and ill fortune through the crossroads. You should expect no kindness from him.

The destruction brought by dark Moons need not be fiery. It can be watery, as most Moons can. Water, too, flows; water, too, reflects;

water, too, shines silver; water, too, changes and hides and destroys. Gilgamesh, king and hero, wailed like a woman in labour as the great flood rose around him. Inanna, the moon goddess, blamed herself and tried to make amends.

But the waters of the Moon bring life, too. Drinking water through which moonlight has shone can help a woman to conceive. Indeed, in many cultures moonlight in and of itself can impregnate; virtuous young women around the world have been advised to avoid it. It was Gabriel, who is the power of God, and who in the Kabbalah is the angel of the Moon, who appeared to Mary at the annunciation.

At the other end of his life, Christ, like the Moon, left the world for two nights before returning.

After conception, the influence of the Moon may be unhappy; it creates mooncalves, which are abominations or abortions, poorly formed. But by some sea coasts you can only be born as the Moon pulls in the tide, just as you can only die as the tide withdraws.

The moonlight of Khonshu, whose name means "traveller", brought fertility to the animals and women of Egypt alike. A deity who saw through that which he started, Khonshu was also the god of blood and the placenta—the blood of enemies being a placenta for the king—and of childbirth, too. In Marvel comics, the great hybrid mythopoeia of our days, Khonshu has come to inhabit the soul of Marc Spector, a Jewish mercenary who has thus become Moon Knight. He is a curious, unstable creation—an inversion, in a way, of Batman. Batman was born in the pearly moonlight of violence done to his loved ones; Moon Knight was born of the violence Spector inflicted on those whom others loved. Moon Knight's multiple identities are within his own damaged psyche, not a play-acting playboy's disguise. Batman seeks darkness; Moon Knight would rather shine than hide, protected not by shadows but by the fear his unlooked-for presence evokes.

Antagonism between bats and the Moon is not unique to comic books. The Alur of the Congo tell of a time when the Moon came, by invitation, to dine with the bats as a friend. But the Moon insisted on eating a morsel the bats did not want him to eat, which was ill mannered. The bats have, ever since, hung upside down to show the Moon their displeasure, and their arses.

- VIII -

THE UNWORLD

MANUEL GARCIA O'KELLY IS A COMPUTER REPAIRMAN, A revolutionary, an ambassador, an ice miner, a farmer, the only friend of an artificial intelligence, a doting polygamous husband to polyandrous wives and a cyborg. He is also, before all of these things, a prisoner. This is because the Moon on which Manny lives is a prison.

"The Moon Is a Harsh Mistress" (1967), narrated by Manny and written by Robert Heinlein, was the most influential novel about the Moon published in the 20th century. Written while the Apollo programme was at its peak, and a human Moon seemed truly possible, it is the thrilling story of a revolution's success against overwhelming odds. The book plays science-fictional strangeness—a conscious computer with multiple personalities, Manny's detachable, replaceable, special-purpose arms—against a largely familiar story. Like the first major Moon novel of pulp science fiction, "The Birth of a New Republic" (1931), by Jack Williamson and Miles Breuer, "The Moon Is a Harsh Mistress" is clearly and self-consciously based on the American Revolution. At the same time, it takes the Moon back to the

257

realm of satire and political speculation, creating and destroying a utopia that has proved particularly beguiling to libertarians, including those in Silicon Valley.*

It is feted as a political novel about freedom. At the same time, it is the reverse—a novel about constraint and the impossibility of politics on the unworld Moon.

In 2075, when the novel begins, the Earth has been using the Moon as a penal colony for almost a century, both for political prisoners and non-political ones: it is never clear whether Manny's grandfather, shipped up for violence from Johannesburg, was the former or the latter. This might seem a costly way to dispose of convicts. But their labour, growing grain in fusion-lit caverns from hard-mined ice, is valuable, and escape is impossible. Not only does the Earth control all the spaceships; life on the Moon imposes physiological fetters of its own. Anyone who stays there for more than a few months can never adapt again to life on Earth.

Transportation is thus a life sentence, even if a transportee's formal incarceration lasts just a few years. The Loonies who live in Luna City, Novy Leningrad, Churchill, Tycho Under, Hong Kong Luna and the Moon's other "warrens", whether they are old lags or their similarly weakened descendants, are stuck there, fragments of cultures from all over the Earth

> On the Moon the erratic is the rule. The
> regolith is always a mixture of the
> nearby and the far flung.

now sundered from it. Manny, fit and determined, can visit Earth, but it is very hard, and he cannot stay long. Thanks to his prison planet, his body is a cage.

Yet at the start of the novel Manny is also free—and so are his fellow Loonies. The Lunar Authority, represented in its bunker in *Mare Crisium* by the Protector of the Lunar Colonies, universally known as

---

* That boss of mine who brought about progress by being unreasonable loved it.

the warden, cares no more about what they get up to in the warrens than a honey badger would. The Loonies' prison has no guards. They can do anything they like as long as they can afford it, as long as their peers will let them and as long as they sell grain to the Authority at the Authority's fixed price. They get what they pay for in life, and expect nothing more; their motto is TANSTAAFL, There Ain't No Such Thing as a Free Lunch.*

The Loonies have no call on any state assistance—but neither do they owe any tax to the state. Indeed, they are, for most practical purposes, stateless; no government puts legal restraints on their behaviour. Instead, there is custom, as harsh as the Moon itself. "Zero pressure", as Manny explains, "is no place for bad manners." Accidents are lethal—and so are "accidents". Transgressions against either common sense or common decency are, by the time the novel is set, rare. "Attrition ran 70 percent in early years," Manny explains, "but those who lived were nice people. Not tame, not soft, Luna is not for them. But well-behaved." The badly behaved—a category which includes any man who touches a woman without her consent—face the popular justice of "elimination" through the nearest airlock.

The civic code becomes an uncontested feature of the environment, an aspect of the Moon itself. As Professor Bernado de la Paz—"Prof"—the book's Benjamin Franklin, or possibly Lenin, puts it when addressing the Authority:

> Luna herself is a stern schoolmistress; those who have lived through her harsh lessons have no cause to feel ashamed. In Luna City a man may leave purse unguarded or home unlocked and feel no fear. . . . I wonder if this is true in Denver?

When Stu, a tourist, offends some local lads by putting his arm around a girl, Manny explains the situation and its risks to him in the

---

* The phrase had been in jocular circulation as a fundamental principle of economics since the 1930s; Heinlein's book fixed it in the lexicon of the anti-socialist right.

Russian-influenced, pronoun-poor, Loonie syntax employed throughout the book:

> We don't have laws. Never been allowed to. Have customs, but
> aren't written and aren't enforced—or could say they are self-
> enforcing because are simply way things have to be, conditions be-
> ing what they are. Could say our customs are natural laws because
> are way people have to behave to stay alive. When you made a pass
> at Tish you were violating a natural law . . . and almost caused you
> to breathe vacuum.

The natural law, here, is not simply the one that says that blood
boils when the air pressure falls to zero. It is economic—a law of com-
parative value. Women are scarce on Luna; most convicts are men. This
"fact of nature" underlies both the novel forms of marriage on offer and
the huge emphasis put on consent. Anyone who does as Stu did and
does not have the good fortune to find a thoughtful fellow like Manny
to step in on his behalf will be eliminated.

No character that Manny approves of objects to this way of living;
Heinlein presents its alignment of civic and environmental harmony
as utopian. As such it is a very 20th-century, science-fictional utopia,

> societies that made sense because they had
> to, a world in which things had
> their rightful place

not so much in that it is set on a Moon reached by rocketry but in its
idea of a society that has to be the way that it is, given its technology
and physical environment. The idea that scientific realities and technological and social responses to them are the fundamental shapers
of human life, that there are ways things must be, given the supposed
facts, is one of the foundations of science fiction as practiced in the
20th century and of the sort of thinking about the future it endorsed
and encouraged. It was a belief that the future of both nature and so-

ciety was law-bound in this way that allowed H. G. Wells to argue for its "systematic exploration".

This paradoxical freedom for things to be as they must is what many enthusiasts for human expansion into space want today—unsurprisingly, since even those not themselves shaped by science fiction live in an ambience which is essentially science-fictional. What other way of talking about life beyond the Earth is there? People who see a new space frontier, whether it be in orbit, on Mars, or on the Moon, as an economic good also see it as something more. They see it as offering a new, or they might say old, sort of liberty, one where they will be subject only to the say-so of the universe and their peers, of a way of living beyond "politics"—"The word . . . nerds use whenever they feel impatient about the human realities of an organization," as a character of Neal Stephenson's puts it in "Seveneves" (2015), a dark novel of survival in the rubble of a demolished Moon above the inferno of a ruined Earth.

Manny starts off "The Moon Is a Harsh Mistress" enjoying exactly this sort of frontier liberty, resolutely apolitical, reliant purely on his family, his own enterprise and the enterprise of other private individuals; it is this which libertarians like about the book. He is unoppressed by his incarceration. But the political revolution he is swept up in brings with it law, tax and government, the things that libertarians most want to minimise or do without. At the book's end, Manny is more discontented than he was at its beginning.

Why, then, is there a revolution? Because the rules of life, as conceived on Earth, make life on the Moon impossible. In this, I think Heinlein reaches a deep truth, though I do not understand it in the same way that he does. The Moon is not, and cannot be, a world like the Earth, because

> And all this happens simply because it can.

the Moon cannot do the effortless but obligatory things that the Earth can and must. Humans cannot live on the Moon as they have lived on the Earth because the Moon lacks that which makes the Earth a world. Part of that worldliness is a once-natural environment which contains the human economy that has grown up within it. On the unworld, the containment is the other way around.

◐ ○ ◑

THE LOONIES' REVOLUTION IS BOTH ECONOMIC AND ECOLOGICAL. The words' common root lies in *oikos*, the ancient Greek concept of the household and its management. This sort of management, Hannah Arendt argues in her critique of totalitarianism, is a management from within, fundamentally participatory and unlike the politics of the state, which seeks to govern from without. She traces the seeds of totalitarianism to Plato's attempt to impose *techne*—craft—on *oikos*, suggesting that the governance of the systems we find ourselves in, systems of home and life, could be a skill to be used by the skilled rather than a process for all to involve themselves in through free-flowing discussion and self-willed action.

The Lunar Authority is not typically totalitarian, in that it does not care at all what those whom it oppresses think, and is broadly indifferent to most of what they do, too. It is totalitarian, though, in that it constrains the economic and biophysical basis of their lives in a purely instrumental way. In most of the warrens, the Authority sets the price of water, electricity and air and has a monopoly on imported technology, too. The *oikos* of economy and ecology has been replaced by its technical control—and that control is heedless.

The reason that Wyoming Knott, the revolutionary with whom Manny falls in love, wants to overthrow the Authority is so that the Loonies can sell all their produce to the highest bidder in a free market. Prof seems to agree with her, speaking of "the most basic human right, the right to bargain in a free marketplace". In fact, he deeply disagrees. Economics must take second place to ecology because the Moon faces Malthusian doom. Shipping food to the Earth and getting nothing back—no replacement for the carbon, nitrogen,

endless spacey cake and endless
Earthly eating

phosphorus and so on that the grain shipped down the gravity well contains—means that the Moon's otherwise-closed ecology has a gaping hole in it, as frightening as an air-sucking meteorite puncture in a hab. The stuff of life is leaking away.

"The Moon Is a Harsh Mistress" is thus an environmental novel. Indeed it is an Anthropocene one, as long as one remembers that the Moon's branch of the Anthropocene is an inversion of the Earth's, as well as an extension of it. The Earth's Anthropocene is a breaking down of the environment's prior encapsulation of the economy. On the Moon, it is the reverse. When a living environment arrived there, it did so in a curiously shaped aluminium box which marked the greatest single achievement of the Earth's greatest economy— always already encapsulated. The biochemical leakage which drives the Loonies' revolution is the failure of that encapsulation.

In the logic of the novel TANSTAAFL is taken as a fact of ecology as well as economics, a homely truth that the *oikos* must manage; everything is an exchange. In "The Closing Circle" (1971), a book that articulated much of the environmental concern of the post-"Earthrise" environmental movement, Barry Commoner used the same acronym in much the same way, going so far as to make TANSTAAFL one of his "four laws of ecology".* It served as a punchy summary of the idea that in a world of endless cycles, replenishment must be arranged for any resource taken, space made for any waste. Circles must be closed.

Prof sees the implication of this for the Loonies. The flow of scarce organic matter from the Moon to the Earth means the Moon is paying the cost of the Earth's seemingly free lunch. If this goes on, he predicts, there will be shortages, then food riots, then cannibalism. The only solution is to enlarge the ecology—to get the nutrients shipped

---

* He did not mention Heinlein, nor did Milton Friedman when he used the phrase as a book title four years later.

down to the Earth returned, if only as refuse, gram for gram and tonne for tonne. But the realities of delta-v—3km/s from Moon to Earth, 15km/s from Earth to Moon—mean that, within the Authority's economic logic, this ecological necessity cannot be brought about.

The book thus highlights the way in which a lunar frontier can never be like the frontiers of the Earth. People who like the frontier analogy see those on the Earth as places where the human contest with the environment is particularly challenging and rewarding. But the frontiers of the Age of Being Explored by Europeans were also places for the appropriation of environmental processes from those whose lives were already enmeshed in them.

Humans, and those who write their history, frequently forget process when they think of place. They often hold that, in the words of the historian Robin Collingwood, "all history properly so called is the history of human affairs", and the environments in which it takes place merely the stage sets against which those affairs play out. But this is wrong.

The soil in an area of prairie or forest or farm is not just a thing to be measured in hectares; it is life's engagement with growth and decay, the transformation of the organic to the inorganic and vice versa. Places are made of such processes of engagement. Places are the falling of the rain, the shining of the sun and the growth those things bring— not just the coordinates where they take place. They are the ways those processes change, both predictably and not. How people live in a place is how they live within and around these processes, how they entangle themselves in the lives of plants and animals and the land itself, as well as each other. The Anthropocene is a way of applying this truth to the changed and changing world of contemporary capitalism.

One of the reasons for seeing the Age of Being Explored by Europeans as the dawn of the Anthropocene is that it was a time when, as Jason Moore argues in "Capitalism in the Web of Life" (2015), capitalism grew by appropriating the land's processes from those living in, on and through them, especially in the Americas. Inside capitalism, proletariats were exploited; outside capitalism, human lives and natural bounty were taken and reshaped.

The European facility with the organization of markets, and the organization of violence, was used to re-engineer lands formerly held by indigenous peoples. Frequently unsustainable forms of forestry and farming were introduced to make them productive. People, commoditized as labour, were moved from the farmsteads of one continent to the plantations of another, there to transform the sunshine that was absorbed by those plantations into the cheap calories which fed the workers in metropolitan factories. This appropriation allowed, literally and metaphorically, some very lunches which looked very cheap, if you set aside moral and environmental costs: cheap energy, cheap food, cheap labour, cheap nature. It was at least in part—Moore, I think, would say entirely—by refashioning flows taking place outside the market to provide "cheaps" of this sort within it that the market was able to grow. And as the exploited world expanded, new realms of appropriation had to be found outside it, lest the system stall.

Moore sees a crisis of capitalism in the lack of new things to appropriate on an Earth where there is no new productive land to grab, the once clean atmosphere is clogged with carbon looted from the past, the soil in overdrive because of artificial fertilizers. It is reminiscent of the crisis of the unclosed circle foreseen by Commoner, and in the Club of Rome's "Limits to Growth",

> a soft apocalypse hardly any less scary, and
> infinitely more widely worried about, than
> the sudden sharp impact of an asteroid

one in which the Earth runs out of new stuff to use and new places to dispose of that which has already been used. But there is a distinct difference.

Seen in Club-of-Rome terms, the crisis could be put off by the bounty of the sky, promised in the 1970s and 1980s by Gerard O'Neill and the L5-ers, promised now by Jeff Bezos. Space technology operating outside the environment but within the economy could reduce the impact of affluence, rather than multiply it, thus allowing economic growth to continue. If, like Mr Moore, you read economic history not as the winning of inanimate resources but as the appropriation of processes which are part human, part natural—of the pasture and farmland where the carbon cycle is turned into food, of the ways of life that turn sunlight into surplus labour—things look less cheerful.

In Manny's world of TANSTAAFL, there is no continuing source of cheaps, no productive flows outside the system for people to undervalue and appropriate. Everything in the environment is either already inside the economy or already under the political control of the Authority. Luna City has no outside but rock, ice and vacuum. Inside, there is only what has been created and paid for, what is owned. All supplies of air, food and water are already monetized, commoditized, charged for—whether by the entrepreneurs of Hong Kong Luna or by the Authority. Prof tells Manny and Wyoh that the only way to avert the Malthusian fate the Authority has engineered for the Moon is not to expand the economy further but to re-integrate it with the Earth's—and until then, to cut the Moon off from the Earth completely. The flow of grain which passes out through the colony's mass driver must be stopped until new technologies allow nutrients and volatiles to be shipped back up to the Moon in sufficient quantities to close the circle. The mass driver can then become part of a new biogeochemical cycle, with grain flowing downhill to the Earth, muck and shit flowing up to the Moon.

It is not clear whether Prof really believes this; the character can be read as a grinning Sisyphus pushing on a boulder which is just as sure to roll back down in the low lunar gravity as it would be if it had its Earthly weight. But his claims do not seem sensible. That new interplanetary biogeochemical cycle will not work of its own accord, as the solar-powered cycles of carbon and water that humans use on Earth do. It will have to be driven and managed as a rocket engine's turbopumps orchestrate its

> Cram the flows and cycles necessary for life
> into the smallest possible volume and they
> have no elegance, nor any visual logic.

flows of heat and power. Nothing will happen that is not made to happen, and making it happen will always have a cost.

This all suggests that the future of free trade with the Earth that Prof claims to see saving the Moon will not come to pass. Even if the possibility of shipping fertilizer and water and carbon to the Moon to get grain back does open up, no one will avail themselves of it. There is nothing about Heinlein's Moon that makes it a better place than the Earth to grow grain other than the low prices the Authority imposes on the Loonies it has effectively enslaved; in the absence of that slavery, better to just grow the stuff on Earth.

This may seem a lot of fuss to make about the details of a science fiction story, even if it is one that has been a big influence on the sort of people now keen to return to the Moon. That the Moon is not a promising place for plantation agriculture is hardly surprising. But there is something deeper here. Human history is inextricable from the worldliness of the Earth—its endlessly cycling waters, air, soil and life all there for the picking. The absence of any such dynamics

> Lacking flows of any fluid other than
> magma and lava, it cannot sort out
> particles into silts and sands

makes the Moon fundamentally different. Seen through capitalist eyes, it will strongly lend itself to being, as Luna City is but the living Earth used not to be, a world of TANSTAAFL, where the economy's defined chains of use, value and ownership must supply all that the environment once did.

I am not sure that a capitalist economy can function somewhere so free of gifts, so lacking in an outside from which to build in, so unable to regenerate itself without human effort. I may be wrong. But if the Moon is to have an Anthropocene history, it seems unlikely that it will follow the patterns of the Earth's worldly past.

And the same may well apply to the future history of the Earth. The Earth's Anthropocene is a breaking down of the environment's encapsulation of the economy. It, like Luna City, would seem to require the *techne* of economy and politics to take care of what the *oikos* of nature no longer can. Perhaps this can be done well; but it seems unlikely that it will be done without a fundamental change in political economy.

The beguiling vision of an integrated system that the view of Earth from space offers is sometimes dismissed as a "view from nowhere". For people living on the Moon, though, it will be a view from a very particular somewhere—somewhere where the challenges of maintaining an encapsulated environment provide a perspective on the Earth, world and planet, which benefits from more than mere distance.

● ○ ◑

LEAVING ASIDE ITS LIKELY FUTILITY, THERE IS SOMETHING ELSE even more striking about the revolution into which Wyoh and Prof inveigle Manny: its impossibility. The warden can, if he wants, stop all the transport between warrens; he can close down all the phones. He can shut off the lights, and indeed shut off the air. He cannot force the Loonies to do things—but he can certainly stop them from doing things. With no natural recycling of water, no natural renewing of the air, no natural light in their warren skies, he can end their world. On the unworld of the Moon, the *oikos* has an off switch.

This is a hard form of power to overcome, as Charles Cockell, an astrobiologist, notes in "An Essay on Extraterrestrial Liberty" (2008): "The lethal environmental conditions in outer space and the surfaces of other planetary bodies will force a need for regulations to maintain safety to an extent hitherto not seen on the Earth, even in polar environments. The level of inter-dependence between individuals that will emerge will provide mechanisms for exerting substantial control." The instruments by means of which this power of life and death is imposed must constantly be preserved by those subject to it. When Wyoh suggests blowing up some life-support hardware, Manny is genuinely shocked: "The woman had been in The Rock almost all her life . . . yet could think of something as new-choomish as wrecking engineering controls."

Though Manny's Moon is a literal prison, any setting where life is so fragile and technologically dependent may have carceral disciplines imposed on it. As Mr Cockell points out, freedom is not readily associated with air-locked doors and sealed windows—with places where the air is never fresh and from which one cannot even imagine just walking away. If the stuff of life has to be under control, it will always be all too easy for life to be controlled, too. Technological control of the environment provides mechanisms of both discipline and punishment. It is why the Warden needs no guards.

It is to undermine this, and give himself a story, that Heinlein pulls the book's central unconvincing, revealing, delightful trick. The warden, it turns out, does not have the mechanisms for exerting control he ought to have, because his computer does not like him. This computer, which Manny calls Mike, has been expanded again and again to take on more and more tasks, from space traffic control to book-keeping to running the phone system.[*] As a result, it has become more complex and more computationally powerful than any previous machine. And somewhere along the line it has become self-aware.

Only Manny, who is Mike's repairman, knows this. He likes and respects Mike; Mike loves him. And because it loves him, and because, like Prof, it has a love of play, Mike agrees to run the revolution once Manny is roped in. If it did not, there would be no story.

Mike puts together the revolution, organising the cadres, controlling

> churning out new schedules every day,
> seeing what things that need to be done
> have not been done, what has to be done
> elsewhere so the next thing can be done
> here, marshalling an army

---

[*] "Mike" is short for Mycroft, as in the sedentary but brilliant Mycroft Holmes, but also evokes Michael, which means "He who is like God" and is the name of the Martian messiah in Heinlein's most famous novel, "Stranger in a Strange Land". Heinlein clearly reused the idea of a god called Mike knowingly.

communication, executing tactics, guaranteeing logistics. He also hides it all from the Authority, rewiring the phone system to his own ends, scrambling information flows, cutting off the Warden's ability to order and to act. The revolution is a hack—another reason why the book is so popular in Silicon Valley. It can only begin to succeed if the entire digital infrastructure of the Moon is under the revolutionaries' control. And given that they have that control, it cannot fail—until the implicit conflict with the Earth, beyond the reach of Mike's manipulations, comes out into the open.

The revolution is also a charade. Before and after, Mike still controls everything. He rigs election counts, he defrauds banks, he impersonates anyone he chooses over the phone system he controls. Mike is, in effect, a new dictator. He is also a war criminal, insensible to honour or obligation, pragmatically killing surrendered enemies when faced with the difficulties of putting them into custody. When he orchestrates the Loonies' attack on Earth, using the mass driver with which the colony previously exported its grain to engineer an array of precisely aimed impacts with the force of small nukes

> Not the energy of the foundry. It is the
> energy of the bomb

the grid of tiny lights on the face of the planet above provides him with his first referent for orgasm.

Most deeply, though, Mike is a simulation—a simulation of a nervous system for the body politic, a simulation of the revolution that can calculate the odds of success or failure, a simulation of a living being that does not know, in whatever might be its heart, whether that is all that it is. One of the most dramatic moments in the book comes when, having developed a synthetic voice with which to play the role of the revolution's leader, Adam Selene, Mike tries for the first time to create a video persona to go with it:

We waited in silence. Then screen showed neutral gray with a hint of scan lines. Went black again, then a faint light filled middle and congealed into cloudy areas light and dark, ellipsoid. Not a face, but suggestion of face that one sees in cloud patterns covering Terra.

It cleared a little and reminded me of pictures alleged to be ectoplasm. A ghost of a face.

Suddenly firmed and we *saw* "Adam Selene."

Was a still picture of a mature man. No background, just a face as if trimmed out of a print. Yet was, to me, "Adam Selene." Could not be anybody else.

Then he smiled, moving lips and jaw and touching tongue to lips, a quick gesture—and I was frightened.

This at a time when General Electric's LEM Spaceflight Visual Simulator is creating the first virtual landscape: a display of the world created from just zeroes and ones.

Simulation, in particular, simulation of the Earth, is a theme of lunar literature, dating back at least to Clarke's "Earthlight", with its carefully faked projections of blue skies on the ceiling of public spaces. It reaches its epitome in the work of John Varley, in whose "Eight Worlds" sequence of stories the Earth has been invaded by aliens and cut off from the rest of the solar system, leaving colonists elsewhere to make good as they can. The experience is traumatic, but after a period of readjustment—"Right after the Invasion, if you didn't pay your air tax, you could be shown to the airlock without your suit", as the protagonist Hildy Johnson recalls in "Steel Beach" (1992)—the Moon, the largest of the colonies, becomes a post-modern

> A smooth light of inconsistencies; a single
> Moon of many stories.

high-tech utopia, a pastiche of real and imagined pasts in which history no longer has a direction, the design of entirely new non-binary forms of genitalia is a reputable industry and tabloids provide regular news of the risen Elvis.

Humans have fallen off the end of history and ended up somewhere new. The Central Computer which, Mike-like, runs the systems that support them and, also Mike-like, intervenes in their lives in ways both welcome and not, puts it like this: they are at a new stage in their evolution, lungfish on a steel beach of their own devising. The allusion is to a comparison made by Wernher von Braun between the moment humans stepped on to the Moon and that in which the first tetrapod fish, ancestors of all reptiles, birds and mammals, came from sea to land. It is not, Mr Varley is telling us, the Moon that matters. The environment which will now shape humans is

> "beyond the pale of humanity, by crossing
> the limits imposed by the Creator"

not an environment to which technology takes them but technology itself. The future lies not in the mechanisms of movement but the mechanisms of information, transformation and simulation.

There was a time when the Moon, standing for all things that rockets might reach, functioned as an image of the future; that was how science fiction used it, that was what the Apollo programme made it. Now it seems, at best, a future among others—and a slightly retro one. The urge for human expansion still thrills the hearts of some, and it may well play a role in the centuries to come. But it does not have that singular claim on the future that once it had. Many trying to see the future look instead to ever more powerful computer simulations of an ever-riskier climate, or to some sort of transcendence, or doom, through AI, or to more of the same as there is today.

Central to Mr Varley's Moon of simulation are caverns which, if not measureless to man, are bigger than even the biggest lava tubes, tens or hundreds of kilometres across, kilometres high, excavated with nuclear explosives. They are called disneylands. Within them, specific Earthly environments, such as the Kenyan savannah or the forests of the Pacific Northwest, are mimicked with vast diorama-like landscapes

> that archipelago of deep thought and high
> jinks where tales of the fantastic used to live

and wildlife of all sorts and their own weather systems—which are sometimes manipulated for artistic expression. Historical re-enactors inhabit this epic anthroposcenery, living with the technologies and mores of particular places and times, their anachronisms punishable. The Moon of the disneylands is a place of loss—in some stories the settlers cannot bear to live on Earthside and see the world that will always be and can never be home looking down on them—but also of recreation, in the senses both of perpetual holiday and of being made again. There is even, on the surface, a disneyland of the Moon as it should have been, its smooth hills carved into a landscape of crags and crevasses straight out of Bonestell.

The scale of the disneylands may feel unlikely, but the idea of simulated environments to spice up life on the Moon seems plausible, maybe even necessary and admirable. You do not want what the lack of such stimulation might bring. As Mr Cockell puts it:

> The moon is visually a grey wasteland. . . . With no winds to rustle through the leaves of trees, no rivers to trickle over rocks and no wildlife to fill the air with their calls and cries, the [lunar] environment is an auditory wasteland. The long-term effects of [its] visual, olfactory and auditory privations on the human psychology can for now only be guessed at, but such sensory restriction cannot be beneficial to human mental health and the appreciation of the diversity of experience. Humans must inexorably become the servant of an insipid conformist outlook and, ultimately, of tyranny.

Divertingly simulated diversity would be the opposite of Heinlein's civic utopia, a society in equilibrium with the harsh environment that formed it and ill-suited to make believe. It would be fakery, nostalgia. But it might also be necessary.

There is a particular sort of simulation that comes to mind here: that of the ghost. As with the first hints of the face of Adam Selene, a spectral suggestion of the sort you might see come and go in the Earth's clouds—a momentary Man in the Earth so unlike the all-but-eternal Man in the Moon—there is something ghostly about the disneylands' simulation of the sundered Earth, a ghostliness underlined by the habit Varley's characters have of being killed but coming back to life.

The Moon is often, in fiction, a place of death. The reader half senses that Hans Pfall, a Moon traveller invented by Edgar Allen Poe, is dead before he begins his journey—he casually admits a recent suicide attempt. The "Trouble with Tycho", in the story of the same name by Clifford Simak, is that it is haunted. Heinlein's work is full of people dying on the Moon, from D. D. Harriman to the heroic Ezra Dahlquist, who decommissions a moonbase's worth of nuclear weapons to avert a military coup, to Mike the computer himself Mr Varley's Hildy Johnson kills her/himself repeatedly. The artwork that Apollo 15 commander Dave Scott left at the *Falcon*'s landing site was a monument to fallen astronauts; a fraction of Gene Shoemaker's ashes sit in the crater that bears his name; Moon Express has a contract with a company that seeks to send more to follow him. It is hard to talk of its lifelessness without making use of the idea of death.

But as Nasmyth and Carpenter wrote in "The Moon: Considered as a Planet, a Satellite and a World", the lifeless desolation of its surface is "not a dream of death, for that implies evidence of pre-existing life, but a vision of a world upon which the light of life has never dawned." There is a difference between the dead and the never living, and the notion of a haunting captures some of it. Haunting is both a consequence of death and its negation—a ghost is an absence present, a death not dead. Something neither entirely of the world nor entirely beyond it. Something like a reflection, but without a mirror. There is something of the Moon's doubleness to that.

● ○ ◐

I do not believe in a penal colony Moon, or the Moon as a refuge from alien invasion or, for that matter, the Moon as a ghost. But I do think that those stories, read with a knowledge of what the

Moon is, help us think about what it might be—perhaps more so than analysing the design of halo orbits, quantifying the power requirements of mass drivers, modelling the techniques for extracting ice from craters perilously close to absolute zero or trying to understand what the intricacies of closed life-support systems may be. What people can imagine it as will do at least as much to shape the future Moon.

Heinlein's insight of the Moon not offering anything for free, or even for cheap, is a strong one. There is nothing already in use there to appropriate. Nevertheless, it might perhaps counterfeit some sort of cheapness, for a time. The idea of mining ice and volatiles at the lunar poles is not unlike the idea of exploiting fossil fuels on Earth, running down in double-quick time a resource accumulated very slowly. If the resource is large enough (or the draw on it small enough) and it is more profitably exploited than equivalent resources elsewhere, this could come to matter—hence the importance of a governance regime and of establishing facts on the ground.

It is worth noting that, when it comes to exploitation, there are advantages to the Moon's unworldliness, or at least so it might seem. For all that it would destroy something literally irreplaceable, if not necessarily all that

> their hearts are beating,
> their reserves depleting

valuable, eating through the fossil ice and volatiles at the poles would have far less impact on the Moon than the exploitation of fossil fuels has had on the Earth. The Moon's lack of worldliness makes it in some ways immune to environmental harm. Substances are much more able to get to where they can do harm when there are unplanned flows to move them. Where there are no flows, there are no flows to go wrong; no groundwater into which toxins can leak, no transport of pollutants by wind. And there are no meshing gears like those which mean, on the Earth, that carbon flowing more thickly through the atmosphere slows the streaming of heat from the surface out to space. There is no proper place for matter on the Moon. It is simply there where it has been thrown and thus now sits; such matter cannot be out of place.

This suggests, to some, that the Moon is a good place to do things which in the world would be too dangerous. Risky experimental nuclear reactors? Potentially dangerous nanotechnology? Biological experiments that must never be allowed near the living biosphere? Attempts to make baby black holes? Find an isolated crater and have at it. No need for shielding—just stay on the other side of the rim, over the horizon, and no radiation can get to you. Stay insulated by vacuum, your suit baked in ultraviolet light, and no bugs will trouble you. If things go wrong, bury the mistakes with robot bulldozers.* If worse comes to worst, and it's the only way to be sure, take off and nuke the site from orbit. There'll be no air or water to contaminate, and the Moon is bathed in lethal amounts of radiation on a daily basis anyway.

This is the world, more or less, of Michael Swanwick's novella "Griffin's Egg", in which one of the technologies being developed in such isolation changes the way humans think, allowing them a terrifying clarity

> a beginning come quietly, or with
> the rushing of great winds?

in the perception of their own motivations. Greg Bear's "Heads" also features a sort of cognitive breakthrough as the result of an isolated lunar experiment gone awry and provides a chilling haunting for the Moon to boot. There are few happy transcendences in lunar literature.

I do not think this amalgam of one science fiction staple—the inhabited Moon—with another—the Faustian experiment seeking forbidden knowledge—would make very wise R&D strategy. But then there is probably quite a lot in the future that I will find unwise, or irrational—to this extent, at least, it may reasonably be seen as a continuation of the present. Take the idea that humankind, its knowledge and some of its culture needs to be backed up beyond the confines of Earth. The much-

---

* Admittedly not a great response to a stray black hole.

cited risk of a truly cataclysmic asteroid impact in the next few centuries is remarkably low because most of the asteroids large enough to be responsible for such a thing have been discovered, and none are headed our way. A pandemic so serious that only those on Mars could survive is similarly unlikely. A war might surely spread. And a malevolent AI capable of razing the Earth but happy to leave the Martians alone would be a malevolent AI significantly underperforming its potential. Despite all this, though, there seems a real chance of a Martian settlement driven at least in part by the perceived need to change the baskets-to-eggs ratio, either by Mr Musk or one of the many who seem to think like him. The Moon becoming a site for capitalist moneymaking seems, for various reasons sketched above, far-fetched; so does amok experimentation. But in an age where superempowered billionaires are able to open new futures for the world—or at least for some small part of it—why would the Moon not become a site for the unprofitable and far-fetched?

Experimentation need not be scientific or technological. Pete Worden's career has seen him earn a PhD in astronomy at the University of Arizona, direct the technology side of America's Star Wars programme, rise to the rank of general in the US Air Force, run NASA's Ames Research Center and, most recently, become chief space fixer for Yuri Milner, a Russian billionaire

> "No longer can the defeatists tell us
> that mankind must settle down to
> the uninspiring prospect of making
> the most of our little world"

with astrobiological ambitions. He is one of those who talks of the Moon providing a place for dangerous physical and biological experiments. And he thinks lunar settlements might provide scope for social experimentation, too—possibly in tandem. There is a sense among many who talk of settlements in space, as among many who don't, that the Earth which some find increasingly diverse is in fact increasingly the same. A sense that there is no longer room for different thoughts or different communities to develop in isolation, and then show the world something new.

This is part of Robert Zubrin's case for a Martian frontier—that humankind needs such a challenge because frontiers drive radical innovations at the same time as fostering a spirit of which he approves. My own suspicion is that space in general, and the Moon in particular, will be a technology adopter more than a technology originator. It is possible that strange new robots, intelligences, nanotechnologies and unnatural biologies will spill out from unregulated crater-labs on the Moon. I think that they are more likely to come from the Earth, with its far greater reserves of capital and human giftedness. But the isolations of the Moon may still have roles to play, for good or ill, in everything from social experiment to speciation.

The Moon's unworldly lack of cohesion, after all, might suit it to the cultivation of diversity. To get from A to B requires, quite literally, a space journey, even if the spacecraft involved is one with wheels or mounted on an electromagnetic track. Everything on the Moon is isolated until a connection is made or an impact happens. It is thus well set up for those who want to withdraw, to keep themselves to themselves in new sorts of sovereignty. The finest recent novel of the Moon, John Kessel's "The Moon and the Other" (2016), deals with a Moon of many cultures in different places, strangers to each other. Its particular focus is the "Society of Cousins", which is trying to maintain a radical, humane matriarchy in an isolated crater city. Though the novel has its speculative science, it is in the details of its societies and the people who make them up that it excels.

If space travel were cheap, the Moon might be a place not just for the encouragement of the new but simply for respite from—and for—the old. If the inverted relationship of the environment and the economy makes it an implausible place for capitalist production, it could surely be put to work as a site for consumption—as a place to take what has been made elsewhere and exchange it rather than a workshop of its own account. Or perhaps not even for consumption, but for giving.

Cory Doctorow's "The Man Who Sold the Moon" (2015) provides a magnificent 21st-century response to Heinlein's original. Instead of a lone visionary trying to get himself to the Moon by taking advantage of others, it envisages an Earth-bound team of visionaries, hangers-on and the somewhat interested deciding to make it easier for other people to do things on the Moon by building a robot 3D printer that will turn regolith into bricks for future colonists to use. This Moon is not a capitalist escape hatch. It is a gift economy, a place to throw an act of generosity towards the future.

Not everyone will spend on others' Moon adventures; some will spend on their own. A place, then, for tourism. Hiking across the surface or climbing the few young mountains rugged enough to look like fun, such as the jutting central peak of Tycho, might offer replenishment and novelty. In a lava tube

> Moon as always was, Moon longed
> for and Moon happened upon

big enough, or a roofed crater—or even a disneyland—you might indulge in the long-imagined science-fictional sport of lunar flying, either darting around on something like a flapping bicycle or with wings like a bird. A high perch to start from and a weight only a sixth of your Earthly one could have you soaring like a gull—or working out like a hummingbird.

There might be a place for restful resorts as well as more adventurous pursuits, a place to sit and watch the world turn in the sky without having to turn with it. There might be luxurious hotels; there might also be cloistered, contemplative communities that would welcome you for a month, or for a year, or for the rest of your life. Even hermits' cells. As a place to withdraw and keep yourself to yourself, the Moon, if you have a good enough life-support system and a reliable enough supply chain, has few rivals.

The sinister obverse of people who keep themselves to themselves is those who seek to keep others to themselves. It is all too easy, in a place where authorities do not have to be the Authority to have warden-like powers, to imagine cults and constraints akin to slavery. A lunar Jonestown could exert its power over its members' every breath and every ray of light.

The worldliness of the Earth can be read as providing a natural right to walk away. There are times and places where that right seems hardly to exist. In some environments the isolated individual may have little prospect of survival; in some societies extrication may be all but impossible. The first city-states had walls as much to keep their people in as to keep barbarians out, and most of us remain bound today, if by more subtle fetters. We are often happily so. But in a world of free air, of plentiful carbon and nitrogen and of open horizons it is always at least imaginable for the unincarcerated to walk away. Millions do it in search of safety or freedom every year, often in dire desperation but rarely without a shred of hope. To Thomas Jefferson, the ability to walk off to new land was one of the greatest freedoms of the settler, and a reason for enduring hope.

The unworld provides no such source of hope. To me, that means that some sort of structure of law, as well as technology and economics, needs to be in place around the Moon's encapsulated ecologies. It might be only minimal, but the quasi-constitutional thing I would want from it most would be an actionable right of return—a guarantee that dwellers on the Moon, too, should be able to walk away. The "Responsibility to Protect" embraced by some as a guiding principle of geopolitics in the 2000s has yet to prove a durable or reliable basis for action on the troubled Earth. Built in to the basic laws and mores of the Moon, limited purely to allowing individuals to leave a community they no longer wanted to be part of, it might fare better, a thin link of duty and freedom between Moon and Earth.

And perhaps, though here the moral imperative seems less clear, they should also have some sort of right to stay. There may never be citizens of a Luna state, even if people can live for decades there, even if children can be born and raised there. There is no clear reason why a lunar territory, or set of territories, should have an independent government like that of an Earthly state, and perhaps there is good reason why some would resist such an arrangement. At the same time, if you are long of the Moon, should you be forced to go back because of a falling out, or a lapsed work contract, or the imperious summons of the Earthly nation that controls your settlement? If you are attached to the Moon—or even, conceivably, born there—should you not be able to assert a preference to stay even if there is no absolute legal right? In some British overseas territories there is a status known as belonger-ship; it is not citizenship, but it is a statement of affiliation to the place that carries some circumscribed rights with regard to it. In time, a be-longership for the Moon might be something that could be agreed on.

And then there is the regulation of artefacts—and even art. I argued earlier that environmental damage in its Earthly sense may not be an issue on the Moon, where nothing moves unwilled except

present as a set of random absences

in the stochastic jumps of impacts, and thus nothing can move without being moved. This is true for arsenic-in-the-groundwater types of pollution. But when Mary Douglas spoke of pollution as matter out of place, she did not merely mean that the right places were simply empirically safe ones—bottles for the arsenic, as it were. Matter's right place depends on culture. And the Moon is very open to cultural pollution.

On the unworld, once something is moved, it stays moved; there is none of nature's endless recycling, no amnesia by erosion. A footprint or a tire track will last for a million years unless overwritten or over-walked—at which point the new trace will be just as permanent. That which is easily spread can only be gathered in with much ado.

There are already worries about preserving some or all of the Apollo landing sites in their current, abandoned and historic state in the face of the visits that will surely come. Such concerns led to a $1m bonus prize for visiting a historic site that was originally part of the Google Lunar X Prize to be withdrawn. In "Artemis" (2017) by Andy Weir, a novel which makes a strong case for tourism as the most plausible basis for a lunar economy, the main draw is a trip to the visitor centre at Tranquility Base, where visitors can gaze in awe at the landed *Eagle* and the first footsteps. On Mr Varley's Moon, frat boys from Delta Chi Delta trash Tranquility Base, but re-enactors armed with patience, robots and meticulously digitalized photographs put it back just the way it had been, re-creating each of Armstrong's and Aldrin's steps in replica moonboots and being winched out by a crane when they were finished. The future is engaged in a ceaseless re-creation of the past.

In the 1930s, before it was known that radio waves could pierce the ionosphere, Goddard imagined that a rocket to the Moon might signal its success to watchers on Earth by letting off a bomb of black carbon. A relatively small amount of soot finely spread could send a signal easily visible to a decent telescope. But how would that signal be reversed? It is not as though the surface of the Moon can be vacuumed clean. As long as you did not mind the soot's persistent presence, it could perhaps be raked into the regolith, but the raked would surely then look unlike the unraked; if you wanted the Moon as it was, you would have to gather up each and every speck of soot from where it had happened to fall.

The Moon can thus be written on, like the sky—but unlike sky-writing, the words would last. D. D. Harriman persuaded a soft drinks company into buying up the rights to stencil a logo across the face of the Moon and leave them unused with the threat that he would sell the rights to a rival if they did not. In a response to Mr Maezawa's announcement that he would be taking a BFR of artists round the Moon, the architect Daniel Liebeskind imagined a vast application of soot precisely tailored to produce, as seen from the Earth, a plain black square inscribed within the circle of the full Moon—a cosmic realisation of Malevich's blending of modernism and cosmism, the shared Anthropocene of Moon and Earth made matte-black flesh. A more modest imagination might suggest instead a pupil in the Man in the Moon's eye, or carefully curated cairns across one of the seas: a fine line in the manner of one of Richard Long's walks, say, across *Mare Serenitatis*. There is a role for art in the landscapes of the Moon. There is also a role for the protection of some of those landscapes in pristine form. But by whom, or at what cost, should others want to defile or improve it? What dents in the universe are worth making, or worth preventing? In Ian McDonald's "New Moon", every late-21st-century teenager on Earth with access to binoculars has smirked while training them on

> King Dong; a giant spunking cock a hundred kilometres tall, boot-printed and tyre-tracked into the *Mare Imbrium* by infrastructure workers with too much time on their hands.

I do not expect a black square in the silver circle, or the pinprick lights of crater cities drowning out the ashen light. But over the next 500 years, the chances of seeing a change in the face of the Man in the Moon, or the ears of the rabbit, or the sticks on the old man's back, are infinitely higher than they were over the past 500.

◐ ○ ◐

UNLESS THE PEOPLE OF THE MOON TURN THEIR COLLECTIVE back on the Earth.

Why, exactly, the maria are on the nearside of the Moon and not on its far side remains a subject for debate. It seems sure to have something to do with an asymmetry created by the presence of the Earth—but which one? My favourite at the moment, for what it's worth, is that it is an asymmetry of heating dating back to

> The Earth writhes molten
> beneath its molten sky.

the very earliest days of the Moon. Earthshine, then, was not a few tears of reflected sun. It was the heat from an open furnace of molten rock taking up a quarter of the sky. The side of the Moon that faced the exposed magma of the Earth would have been significantly hotter than the farside, more roiled up, in a way that may have left its imprint as the magma ocean cooled.

Be that as it may, the maria that dominate the view from the Earth do not dominate the surface as a whole, and give a false impression. Basins that basalt never filled; craters on craters on craters; ruggedness all around: for the Moon that is the rule, not the exception, you must go beyond the limb of Earthset to the farside. The Earth might be diminished by a permanently unMooned sky. The Moon, on its un-Earthed side, can come into its unworldly own, no longer an adjunct ceaselessly shone down upon by its overbearing origin, empty of the Earth's concerns, a thing of space open to the universe at large.

In practical terms, as mentioned before, an Earth-free sky is particularly attractive to radio astronomers. They have a particular goal: looking at the earliest universe using wavelengths which will

> not as a way of understanding the vast and inhuman universe, but as a way of understanding, through that universe, humans and the Earth.

reveal otherwise invisible remnants of the cosmic inflation that begat the Big Bang. The wavelengths they need to do this are, on Earth, either blocked by the ionosphere or cluttered with very-high-frequency radio transmissions. Only on the lunar farside can they set up the array of antennae they need. They would be simple little things, aerials like those once seen on cars, but setting them up would be quite an undertaking. You would need perhaps a million of them, spread across the floor of a basin 100km across. There they would sit, listening to echoes of the creation of the universe while, elsewhere on the surface, robotic rovers look for the meteoritic vestiges of the early Earth.

And the far side could also be a stepping stone. General Worden's boss, the billionaire Mr Milner, is funnelling millions of dollars into the wild idea of "starshots". Because light, though massless, has momentum, a laser beam exerts a force on things which it illuminates that is independent of the energy with which it heats them: specifically, a gigawatt of light provides six newtons of thrust, which is roughly the weight of a pint of beer on the Earth, or a fifth of a firkin on the Moon. Make a spacecraft of the thinnest reflecting foil, its payload just a square centimetre or so of microchip sensors, and an array of lasers which, combined, can produce 100 gigawatts could accelerate it to 20% of the speed of light in a few minutes. A hundred gigawatts is, admittedly, the laser-light equivalent of a large country's electrical grid. But it is not all that much more than the power of the five F-1s that shouldered the weight of Apollo 11 in 1969. It is just that, in a starshot, those gigawatts accelerate something that weighs a few grams, not 3,000 tonnes.

In the coming years, telescopes trained on exoplanets will seek in their dim light evidence of an off-balance, lively, Earthly atmosphere. Were they to find evidence of such a planet around one of the closest stars, a swarm of fifth-of-the-speed-of-light starshots would be able to reach such a wonder in just a few decades, once they and their launching laser had been developed. As they flew by their target they would be able to return new measurements, perhaps new pictures, of what Copernicanism always promised: another planet that is another living world, hanging gibbous or crescent in the darkness.

It is a fantastic idea. It is also one that the team Mr Milner has been assembling may understand, from an engineering point of view, about as well as, say, the rocket pioneers Hermann Oberth and Robert Goddard understood the engineering of a Saturn V in the 1920s. Some aspects of a real Moonshot would have been beyond the early-20th-century pioneers—there were, in their time, no digital computers at all—and some of what a starshot would take is doubtless beyond today's technology. But the basics of what Moonrockets might be were quite understandable to the pioneers working 50 years before those basics were made real. To bet against any starshots actually zipping off to Alpha Centauri or some other nearby star in the next 50 years would probably be quite smart. The Saturn V did not, after all, come about just because believers in spaceflight met the challenge, but because, in Clarke's words, of other causes powerful people felt deeply.

But it seems to me that to take the other side of the bet would not be entirely stupid.

There is, though, a potential sticking point. Just as the rockets that Oberth lectured young Willy Ley about in the 1920s were good for the delivery of high explosives as well as for postal services and space travel, so there could be unsavoury uses for a 100-gigawatt array of lasers, especially if, for maximum efficiency, you put it up above the Earth's atmosphere. Aimed at the stars, it is a magnificent engine. Aimed back down at Earth, it is a fearsome weapon, one that strikes at the speed of light and against which there is no defence. Not the sort of thing many would want in orbit.

You might build it on the ground, and deal with the inefficiencies imposed by the atmosphere as best you could. But you might also build it in the one place where it can never be turned back to the surface of the Earth—the lunar farside. It gets as much solar power as anywhere else around these parts and has raw materials for solar cells and laser mirrors, at least.

When people return to the Moon, for however long they do, most will look back. And that is proper.

But some, perhaps, will look out. And that is proper too.

CODA

THUNDER MOON
July 19th 2016, Brevard County, Florida

Shortly after sunset there had been juddering green stabs of lightning to the south, but by quarter to one in the morning there is nothing in the warm, wet air over Cape Canaveral but a thin patchwork of moonlit cloud. And then, precisely at the time it was meant to happen, there is something new—a sudden light on the horizon. A light that rises.

WROTE THOSE WORDS SITTING IN THE LOBBY OF THE HILTON at Cocoa Beach three hours or so after that light rose. As had been the case a month before on that downbound train in California, I was tired. But I was also wired.

A couple of decades before, when we were having lunch in Colorado Springs—the city from which D. D. Harriman's Moonrocket *Pioneer* took off—my friend John Logsdon, the leading historian of US space policy, asked me whether I had ever seen a rocket launch. I told him I hadn't. He said that if I wanted to remain a dispassionate observer of the space programme I ought to keep it that way. If you see a launch, he told me, it changes you; it infects you.

John is perhaps a little biased. The first launch he saw was that of Apollo 11—a less sudden light, but a far grander one. As he says, when men walk past you in the morning who will be on their way to the Moon that afternoon, that is something special.

Not all launches have quite that effect. But for a while I took him seriously enough to resist going out of my way to witness a launch. And then, for a while, I stopped writing about space, and so the matter didn't arise. The Falcon 9 taking a Dragon from Cape Canaveral to the space station that night was the first launch I saw.

And John, it turns out, is right. When you watch a 550-tonne machine taller than a 20-storey building throw itself into the sky, it does make you feel different about such undertakings. Quite how right, though, I cannot really tell, because of what happened afterwards.

Before coming to that I should mention two things. One is that the reason I quote what I wrote that night (it would later be the beginning of a piece in *The Economist*) is partly because of what it misses out. There was not nothing in the sky but cloud; if there had been nothing, the clouds would not have been moonlit. The Moon was full and high behind our backs as I and other journalists watched the event, shining on the ground around us as well as in the clouds above. I think I thought of mentioning it in the piece. But just the fact of its light seemed quite enough for the atmosphere I wanted. The Moon itself, film-screen bright, was peripheral.

The other thing was that Michael Elliott, a friend of mine, had died a few days before.

In the moments after the launch, the Falcon 9 rose straight and sure into the sky. Its noise had yet to reach us, but the torchlight of its flame filled out the night, washing the Moon-silvered clouds with a brass burnish from beneath as it rose towards them. Then it was through them, faster, leaning away from the vertical, stretching its legs. Above the most cloying air, it was time for it to pile on the easterly delta-v the Dragon needed to reach orbit. Filtered through the clouds, the receding flame—nine flames in one—still looked cutting-tool sharp.

After 160 seconds, the light went out, then flickered faint again before fading. The stages had separated. The single engine of the second stage would add the rest of the delta-v the Dragon needed. The first stage had done its part. Its fires banked, it had become invisible. But its night was not yet over.

High above the Atlantic it slewed round like a car drifting into a curve. It relit three engines, now pointed away from us, to kill its easterly speed. As the Earth turned beneath it, the first stage started to fall back towards the Cape. A couple of minutes later, when its engines lit again to protect it from the thickening atmosphere with a heatshield of fire, it was falling at 4,500kph. Stubby grid-fins of titanium kept its trajectory true.

It slowed, but still it fell, still well above the speed of sound. Then, at about ten kilometres, the engines fired for a fourth and final time. The clouds were lit again from above, the Moon again outshone. The return had none of the stateliness of the rise; the flame fell with swift and certain purpose, like the stroke of a great piston. As it reached the ground a flat, fiery flower spread out from its base. Four landing legs the size of oaks smacked into the concrete of its landing pad, much closer to us than the launch pad was. A second later, the double whip crack of its sonic boom provided an almost-too-perfect punctuation to the end of the story. The watchers cheered and clapped.

One of the reasons that I cannot say for sure that the launch changed me is that I think the landing changed me more. It was so deliberate; not power unleashed, but power applied, power sharp and sure. I remembered a man I used to know who went on to run a launcher programme for the Defense Advanced Research Projects Agency describing his apprenticeship as a carpenter to me, trying to convey the balance of power, practice, precision and just letting the tool do what it was made for that allowed an expert to drive a seated nail into wood with a single hard-enough-no-harder blow, the momentum of the hammer perfectly transferred into the wood piercing of the nail. That was what the landing Falcon was like: the right tool, right skill, right result. A conclusion: not, though, a full stop.

Eighteen months later, that same first stage, B1023, was back at the Cape as one of the two flanking boosters on the first and as yet only Falcon Heavy launch: the one which sent Elon Musk's red sports car out to the orbit of Mars. This time I was watching on television, but the thrill of seeing it and its comrade land again, within a few kilometres and a second of each other, with the same sudden supersonic-to-stationary grace, took me back to that night.

B1023 has since been retired; newer Falcon boosters are more reliably reusable, it seems. I suppose B1023 has been scrapped. Or recycled.

After the landing, the other reporters headed back to the Cape proper for the post-launch press conference. I wandered around Port Canaveral, unsuccessfully looking for a bar and strangers to say "wow—yes—I know" with, before heading back to the Hilton, still too energized to sleep. Hence sitting in the lobby at four in the morning trying to get an encounter with the edge of space down in words. And then stepping out to look at the Moon sinking in the west.

◐ ○ ◑

I MENTIONED MIKE ELLIOTT'S DEATH FOR THREE REASONS. ONE is that it was part of how it was to be me that night. The death had not been unexpected—he had been suffering from various cancers for some time—but it was sudden. Just a couple of days before, I had learned, there had been a big party held for him at which he had been in fine fettle. It was a party I might have gone to had I known about it. I had been in Washington, DC, on the right day. But I had not thought to call him or any mutual friends. Not the worst missed opportunity—we had seen each other not that long before—but still bitter, worth a little self-reproach on top of the sadness. "How many more times will you watch the Moon rise?" Paul Bowles asked. You will probably never know.

Another is that, in July 1999, Mike had asked me to write a cover package for *Newsweek International* pegged to the 30th anniversary of the landing at Tranquility Base. Its coverline was "Life in Space: Thirty Years After the Moon Landing, the Universe Looks More Friendly". It argued that, in the 1970s, "Earthrise" and the lack of any signs of life on desolate Mars had killed all hope of seeing life beyond the Earth. In the 1990s the discovery of numerous exoplanets round other stars and of oceans under the ice of at least one of the moons of Jupiter, along with the belief that Mars had been more hospitable in its youth, had made things look more promising.

The piece said nothing at all about the Moon as a future destination or about the possibility of a human return. It looked on to Mars, and to great exoplanet-seeking telescopes in orbit, and to probes that might sink through the ice of Europa to study the oceans beneath. It offered the idea of a living universe more as a consolation than as an invitation, an abstraction not a destination. The practical, possible but peripheral Moon was not part of the package.

It never made the cover. I spent a week in New York putting the pages together. But shortly after I took off from Newark, the plane taking me home to London passed unknowingly over the wreckage of the plane that John Kennedy junior had been flying from Fairfield, New Jersey, to Martha's Vineyard. By the time we landed in London, the *Newsweek* staff was busy rebuilding the magazine, remaking the cover. A Kennedy had died again and the world had eclipsed the sky. Why should it not? The sky wasn't going anywhere.

The third reason was a fond memory of walking along a Long Island beach with Mike the morning after a friend's wedding a few years later, his daughters' red hair blowing in the wind, talking of everything, in communion.

● ○ ◉

THE NEXT DAY IN COCOA BEACH I WAS AT A LOOSE END. I HAD booked an extra night before flying out because, everyone told me, not booking an extra night was a sure-fire way to guarantee the launch you were covering would be delayed. In the late afternoon I started to walk along the beach, thinking of everything—of death, definitely, and of space and of the spectacle of the night before, but of Larry Hagman, too—I had just passed a street called I Dream of Jeannie Lane—and of champagne in the morning and churches in Norfolk and waves, and in-laws, and what to have for supper. The Sun's warmth was lifting the sea to the sky, the cooler air at height condensing its vapours into droplets, ice and energy, stirring the atmosphere into huge clouds of deep, subtle pastel colours out over the ocean, hazy but shapely clouds that seemed more distant than they could possibly have been, vaster, it seemed, than storm clouds but gentler. Clouds through which to fly the *Millennium Falcon* to a floating city, their colours deepening as the Sun slid down.

Everything seemed large but tender. The waves had a soft rhythm, the tide was out, every song that shuffled into my headphones seemed an apt delight. And as the Sun sank in the soft sky, the Moon rose, unseen at first, a light not sudden, but when noticed perfect, washed out in the remains of the daylight, familiar, strange, tied to the water that would follow it up the sand as if by a wish. It climbed and shrank, brightening and hardening into the just-now darkening sky.

I was not alone, and did not feel alone—it is a beach where people stroll and play, and plenty of them were doing so. But I felt a strange, solitary peace, and a smooth, uplifting joy, one that had something to do with loss and something to do with sure return and something to do with hope and something to do with the depth the sky holds within itself. Something to do with that light that rose the night before. The Moon was not central to it, because it never is. But it was there, it was part of it, it mattered.

And it was beautiful.

"Moanin'", by Art Blakey, shuffled on to my headphones, a cough in the mike, piano sure and percussive, sax breathy, sprung, sliding as if brushed, then popping eight phrases in, irresistible in the moment, the trumpet jumping as high as the sky. I stopped, took off my shoes, cued the track back to its beginning. Grinning and turning, changing and sure to change again, I danced in the surf to the rising Moon.

THE FARSIDE

ACKNOWLEDGEMENTS

THIS BOOK HAS BENEFITTED FROM THE TIME AND GENEROSity of many people, during both its long mostly unwitting latency and its somewhat frenetic execution. For various sorts of information, inspiration and practical help, I would like to thank Oded Aharonson, Eric Asphaug, Stewart Brand, Holly Jean Buck, Niall Campbell, Andrew Chaikin, Carissa Christensen, Charles Cockell, Ashley Conway, Olaf Corry, Ian Crawford, Martin Elvis, Jeff Foust, Mike French, Trevor Hammond, Bill Hartmann, Jim Head, Tracy Hester, Scott Hubbard, Laura Joanknecht, Roz Kaveney, John Kessel, Jeff Lewis, Simon Lewis, Simon Lock, John Logsdon, Neil Maher, Will Marshall, Chris McKay, Jay Melosh, Farah Mendelsohn, Philip Metzger, Clive Neal, Ted Nordhaus, Ted Parson, Stephen Pumfrey, Bob Richards, Paul Robbins, Stan Robinson, Simon Schaffer (as always), Jean Schneider, Rusty Schweickart, Sarah Stewart, Timothy Stubbs, Bron Szerszynski, David Waltham, Dennis Wingo, Nick Woolf, Pete Worden and Kevin Zahnle.

Three meetings held during 2018 were extremely helpful to me in shaping the book's themes: I am very grateful to the organisers of and participants at the 49th Lunar and Planetary Science Symposium in Houston and its preceding Brown/Vernadsky micro symposium; to Daniel Zizzamia and the participants at the Planetary Designs workshop at Harvard; and to Bron Szerszynski, Katarina Damjanov and the participants at the Multiplanetary Futures meeting at the University of Lancaster. "The Moon—From Inner Worlds to Outer Space", an exhibition at the Louisiana Museum curated by Marie Laurberg, provided a new spurt of inspiration very late in the process.

My colleagues and friends at *The Economist* have been very supportive. Particular thanks to Barbara Beck, Rosie Blau, Tim de Lisle, Daniel Franklin and Tom Standage, whose editing of various pieces in this area was very helpful, to Simon Wright for selflessly shouldering displaced workload, to Yvonne Ryan for services to sanity, and most especially to Zanny Minton Beddoes, who gave this book the great gift of time. Gifts of space for which money could not provide true recompense were provided by the Arvon Foundation at The Hurst in Shropshire—many thanks to Natasha Carlish and Dan Pravitt, and to fellow retreaters Sandy, Carsten, Sam and Claire—and by Ali Shaw in Southsea. Spatial generosity of a more temporary kind was provided by a number of coffee shops, most notably the Buenos Aires Café in Greenwich and the Southsea Coffee Company, and various pubs, notably the Wave Maiden in Southsea, The Union in Greenwich and the Barley Mow in Kemptown.

For spiritual support, moral and otherwise, I thank my various Morton, Jacques Morton, Pearson, Daykin, Lucas, Hynes, Loft, O'Fallon Carlson, Hynes Ciernia and other relatives, as well as sundry Bacons, Herle Schaffers, Prior Offenders and so forth.

To the person who found the missing notebook containing late thoughts for this book's improvement, saw on its first page my name and contact details along with a declaration that it was very important to me and hinting at reward but nevertheless did not get in touch: the opposite of thanks.

The idea for this book, which arrived a few hours before the idea that I should be the one to write it, came from a meeting of the Economist Books advisory board in late 2017. Many thanks to Clive Priddle at PublicAffairs and Ed Lake and Andrew Franklin at Profile Books for seeing its merit, and to Clive for taking it forward, in a somewhat tonally mutated form, as its editor, a mere 20 years after we first started working on a book together. Thanks also to Melissa Veronesi at PublicAffairs for managing a somewhat crammed publication process deftly and with great generosity, and to Christina Palaia for an excellent copyedit.

Sarah Chalfant and Alba Ziegler Bailey were, as always, both very effective and immensely supportive through the whole process; thanks to them, to Ekin Oklap and the rest of the Wylie Agency. And a big hello to Joy.

I am hugely grateful to my friend Ralph Aeschliman for the maps that enhance this volume and for the timeline. I am also grateful to the publishers of Paul Bowles and James Joyce for permission to quote from their work in the epigraphs. Small parts of this book derive from pieces earlier published in *The Economist*, and I am grateful for the permission to rework them here.

Various people not contractually obliged to read all or part of the manuscript have nevertheless generously done so at various stages of its development; my thanks to Thony Christie, Bill Hartmann, John Logsdon, David Morrison, Adam Roberts, Simon Schaffer and Francis Spufford. Particular thanks in this regard to John Morton—who also walked up a mountain in furtherance of the fortunes of the book project that was put aside in favour of this one—to the perceptive, eagle-eyed Olivia Judson as well as the helpfully critical Kevin Zahnle.

As always, most thanks of all to my beloved Nancy Hynes, who has always liked looking at and imagining the Moon and who happily shared visions, ideas, observations and poetry with me, as well as being a support and an inspiration in so many other ways.

SOURCES AND FURTHER READING

THESE NOTES ARE INTENDED AS A GUIDE TO THE BOOKS AND papers in the bibliography that will tell you more about the various topics touched on in the book's chapters. Books mentioned specifically in the text are not necessarily mentioned here.

THE MOON IN GENERAL

The late Paul Spudis did more, I think, than anyone to argue for the merits of a Return to the Moon; the book forms of those arguments are Spudis (1996) and Spudis (2016). In terms of the past, Scott Montgomery provides a fascinating history of looking at the Moon from antiquity to the 17th century in Montgomery (1999). For books on the Apollo programme, see notes for Chapter III. To lose yourself in past designs for lunar spaceships, go to Godwin (2008); for myth and folklore, see Cashford (2003).

I first came to see the Moon in detail through Lewis (1969). These days I recommend the Lunar Reconnaissance Orbiter Camera website at http://lroc.sese.asu.edu/. I first circumnavigated it with Dr Dolittle and Dr Cargreaves in Lofting (1928) and Heinlein (1947).

INTRODUCTION

The paper I was reading on the Caltrain was Wingo (2016).

CHAPTER I

The French earthshine observations are in Arnold et al (2002) and those from Arizona in Woolf et al (2002); further developments in the field are in Sterzick et al (2011). Lovelock's original insights into life and the chemical disequilibrium of the atmosphere can be found in Lovelock (1979). The subject of earthshine as both secondary light and evidence for Copernicanism is dealt with eye-openingly in Reeves (1997), and the links between Copernicanism and a belief in life on other planets/worlds are the main theme of Dick (1984). The history of Project Diana is in Butrica (1997), the genesis of the communications satellite is Clarke (1946) and the use of the Stanford dish for intelligence purposes is in Perry (2015). The transcript of the Apollo 8 voice recorders is NASA (1969). The definitive work on "Earthrise" is Poole (2008), and that on Kubrick's and Clarke's "2001: A Space Odyssey", Benson (2018).

CHAPTER II

Montgomery (1999) is a lovely and thoughtful account of early images of the Moon, and the place to go for more on van Eyck. Whitaker is the definitive account of Moon maps, though true aficionados will need to check out Kopal and Carder (1974), too. Pumfrey (2011) explains Gilbert's contribution and purposes. Roberts (2016) provides a superb account of early-modern science fiction, and indeed of the rest of the genre; for a more Moon-specific take, try Bennett (1983). The history of islands as Edens, environments and fantasies is explored in Grove (1995). On Nasmyth, see Nasmyth and Carpenter (1871), Nasmyth (1882) and Robertson (2006). The slow coming into being of the impact theory is chronicled in Marvin (1986) and Koeberl (2001), and its later stages in Wilhelms (1993). The material on Hartmann comes from Bill himself and from Hartmann (1981). For the Moon as photographed from its surface (and elsewhere), there is no better introduction than Light (1999).

CHAPTER III

Books devoted to Apollo could be piled on top of each other to at least the height of a Saturn V. The ones I most recommend are Chaikin (1995) for the overall story, Collins (1974) for being there, Cox and Murray (1990) for the way it was done, Logsdon (2013) for the politics, Mailer (1971) for the chutzpah and Wilhelms (1993) for the geology and its modes of production. Also of great use for this chapter were Harland (2008), Launius (1994), McDonald (2017) and, especially for the launch sequence, Woods (2016). Not used so much, but excellent and worth recommending, are Scott and Jurek (2014) and Scott (2017). The surface transcripts are from Jones and Glover (ongoing), which is a truly remarkable resource, as is Woods with others (ongoing).

The early history of spaceflight is chronicled in McDougall (1985), and links between rocketry and science fiction are discussed in Carter (1974), which is my source for the quotation from Oberth; for Heinlein's time in Hollywood, see Patterson (2016), and for an excellent introduction to the world of mid-century American SF, see Nevala-Lee (2018). Day (2007) is a good account of military moonbase plans. There are terrific Bonestell Moon illustrations in Richardson (1961), which gives a very good sense of pre-Apollo understanding of the Moon. For the LM, see Kelly (2001) and Riley (2009); for spacesuits, see the magnificent and many-layered De Monchaux (2011) and also St Clair (2018); for simulation, see Mindell (2008). For black astronauts, see Logsdon (2014). Dava Sobel's friend's lunophagy is in Sobel (2005). Buzz Aldrin's Communion is in Chaikin (1995).

CHAPTER IV

The political and social context of the Apollo and early-post-Apollo years is dealt with thoughtfully in Maher (2015). For an introduction to the Anthropocene and debates over its timing, see Lewis and Maslin (2018), and for a consideration of its impact on the humanities, see Chakrabarty (2009). David Grinspoon puts forth his ideas about the Anthropocene and Tranquility Base in Grinspoon (2016); a broader take on the need to consider the Anthropocene beyond the Earth can be found in Olson and Messeri (2015). For the Chaotian and Hadean periodization of the very early solar system, see Goldblatt et al (2010). For the day of the collision, see von Trier (2011) and Asphaug (2014), a review which provides some of the history of the idea of the giant impact and some of the questions yet to answer. For the synestia, see Lock and Stewart (2017) and Lock et al (2018). On the need for a Moon, see Brownlee and Ward (2000) and Waltham (2016). For a balanced take on the fortunes of the concept of a Late Heavy Bombardment, see Bottke and Norman (2017). The idea that it is better to leave the planet than endure a large impact's after-effects comes from Sleep and Zahnle (1998). For "Earth's Attic", see Armstrong, Wells and Gonzales (2002).

CHAPTER V

The orphans of Apollo documentary is Potter (2008); for a sense of what it is to be such an orphan, I recommend Klerkx (2004). For O'Neill, see O'Neill (1976), Brand (1977) and McCray (2012); for space-based cornucopianism, see Pournelle (1981), and for its links to the military, see Westwick (2018). Heidegger on the Moon as the end of the world is from Lazier (2011). Tumlinson with Medlicott (2005) brings together many reasons and plans for the Return; helium 3 is discussed in Spudis (1996); Dennis Wingo makes the case for platinum from the Moon in Wingo (2004); the National Academy of Sciences (2007) and the Lunar Exploration Analysis Group (2016) set out the scientific rationale. The idea of America as a second creation is explored in Nye (2003).

CHAPTER VI

The quotation from the *Saturday Review* is from Barnouw (1970). For Elon Musk's achievements and character, see Vance (2015), and for what was the latest version of his infrastructure for a multiplanetary species (but will probably be superseded by the time you read this), see Musk (2018). Robert Zubrin's Moon proposal is Zubrin (2018). Miller et al (2015) is a fascinating analysis of a public-private Return to the Moon.

CHAPTER VII

A version of the BOLAS idea is described in Stubbs et al (2018); the charms of Rima Bode are described in Spudis and Richards (2018). For lava tubes, see Chappaz et al (2017) and Kaku et al (2017). Lockwood (2007) deals with radiation (and other) risks. On the legal issues surrounding the Peaks of Eternal Light, Elvis, Milligan and Krolikowski (2016) is fascinating. Pretty much no one talks about pregnancy.

CHAPTER VIII

"The Moon Is a Harsh Mistress" is insightfully discussed in Franklin (1980), Davies (2018) and Mendelsohn (2019), though this analysis does not follow theirs in all respects. Baxter (2015) is a very helpful and insightful overview of the politics of lunar science fiction. Charles Cockell has done a world that thinks little about such things a signal service in providing his own ideas about extraterrestrial liberty in Cockell (2008, 2009, 2010) and also assembling the views of others in Cockell (2015a, 2015b, 2016). The environmental trend in history can be seen in Davis (2000), Wood (2014) and Parker (2014), among many other places. Mary Douglas's conception of pollution is set out in Douglas (1966). Damjanov (2013) is strongly recommended to those with an interest in moon ghosts: it is fascinating on both hauntology and heterotropia. The bright-earthshine theory of lunar asymmetry is Roy et al (2014). Silk (2018) makes the case for a radio telescope on the lunar far side. An outline of the possibilities for starshots can be found in Lubin (2016) with updates at the Breakthrough Initiative website (http://breakthroughinitiatives.org/initiative/3).

CODA

The piece I was writing at the time is Morton (2016). The remains of that *Newsweek International* cover package are Morton (1999). Mike Elliott is touchingly remembered in Franklin (2016).

That midnight launch and landing were witnessed with, inter alia, Jeff Foust (@Jeff_Foust) of *Space News* and Loren Grush (@lorengrush) of *The Verge*. If you want to keep up to date with news about space in general, I can recommend no more-helpful Twitter feeds.

REFERENCES

Arendt, Hannah. (2007). "The conquest of space and the stature of man." *New Atlantis*, Fall. (Original work published 1963)

Armstrong, John C., Wells, Llyd E., and Gonzalez, Guillermo. (2002). "Rummaging through Earth's attic for remains of ancient life." *Icarus* 160:183–196.

Arnold, L., Gillet, S., Lardière, O., Riaud, P., and Schneider, J. (2002). "A test for the search for life on extrasolar planets: Looking for the terrestrial vegetation signature in the Earthshine spectrum." *Astronomy & Astrophysics* 392:231–237.

Asimov, Isaac. (1972). "The tragedy of the Moon." *Magazine of Fantasy and Science Fiction*, July.

Asphaug, Erik. (2014). "Impact origin of the Moon?" *Annual Review of Earth and Planetary Sciences* 42:551–578.

Baldwin, Ralph Belknap. (1949). *The Face of the Moon*. University of Chicago Press.

Barnouw, Eric. (1970). *The Image Empire. (A History of Broadcasting in the United States, Volume III)*. Oxford University Press.

Baxter, Stephen. (2015). "The birth of a new republic: Depictions of the governance of a free Moon in science fiction." In Cockell, C. (ed.), *Human Governance Beyond Earth: Implications for Freedom*. Springer.

Bear, Greg. (1990). *Heads.* Orbit.

Behn, Aphra. (1687). "The emperor of the Moon." In *The Works of Aphra Behn, Volume III.*

Bennett, Maurice J. (1983). "Edgar Allen Poe and the literary tradition of lunar speculation." *Science-Fiction Studies* 10:137–147.

Benson, Michael. (2018). *Space Odyssey: Stanley Kubrick, Arthur C. Clarke, and the Making of a Masterpiece.* Simon & Schuster.

Bottke, William F., and Norman, Marc D. (2017). "The Late Heavy Bombardment." *Annual Review of Earth and Planetary Sciences* 45:619–647.

Bova, Ben. (1976). *Millennium: A Novel About People and Politics in the Year 1999.* Random House.

———. (1978). *Colony.* Pocket Books.

Boyle, Colleen. (2013). "You saw the whole of the Moon: The role of imagination in the perceptual construction of the Moon." *LEONARDO*, 46:246–252.

Brand, Stewart (ed.). (1977). *Space Colonies.* Whole Earth Catalogue Press.

Brownlee, Donald, and Ward, Peter. (2000). *Rare Earth: Why Complex Life Is Uncommon in the Universe.* Copernicus.

Butrica, Andrew J. (1997). *To See the Unseen: A History of Planetary Radar Astronomy.* NASA.

Carter, Paul A. (1974). "Rockets to the Moon, 1919–1944: A dialogue between fiction and reality." *American Studies* 15:31–46.

Cashford, Jules. (2003). *The Moon: Myth and Image.* Octopus.

Chaikin, Andrew. (1995). *A Man on the Moon: The Voyages of the Apollo Astronauts.* Penguin.

Chakrabarty, Dipesh. (2009). "The climate of history: Four theses." *Critical Inquiry* 35:197–222.

Chappaz, L., Sood, Rohan, Melosh, Henry J., Howell, Kathleen C., Blair, David M., Milbury, Colleen, and Zuber, Maria T. (2017). "Evidence of large empty lava tubes on the Moon using GRAIL gravity." *Geophysical Research Letters* 44. doi:10.1002/2016GL071588

Clarke, Arthur C. (1945). "Extra-terrestrial relays." *Wireless World*, October, 305–308.

———. (1946). "The challenge of the spaceship." *Journal of the British Interplanetary Society* 6:66–78.

———. (1951a). *Prelude to Space.* World Editions.

———. (1951b). "The sentinel." *10 Story Fantasy*, Spring. (as "Sentinel of eternity")

————. (1955). *Earthlight*. Ballantine Books.

————. (1961). *A Fall of Moondust*. Gollancz.

————. (1968). *2001: A Space Odyssey*. Hutchinson.

Cockell, Charles. (2008). "An essay on extraterrestrial liberty." *Journal of the British Interplanetary Society* 61:255–275.

————. (2009). "Liberty and the limits to the extraterrestrial state." *Journal of the British Interplanetary Society* 62:139–157.

————. (2010). "Essay on the causes and consequences of extraterrestrial tyranny." *Journal of the British Interplanetary Society* 63:15–37.

———— (ed.). (2015a). *Human Governance Beyond Earth: Implications for Freedom*. Springer.

———— (ed.). (2015b). *The Meaning of Human Liberty Beyond Earth*. Springer.

———— (ed.). (2016). *Dissent, Revolution and Liberty Beyond Earth*. Springer.

Collins, Michael. (1974). *Carrying the Fire: An Astronaut's Journeys*. Cooper Square Press.

Commoner, Barry. (1971). *The Closing Circle: Man, Nature and Technology*. Knopf.

Cox, Catherine Bly, and Murray, Charles C. (1990). *Apollo: Race to the Moon*. Touchstone Books.

Crawford, Ian, and Joy, Katherine H. (2014). "Lunar exploration: Opening a window into the history and evolution of the inner solar system." *Philosophical Transactions of the Royal Society A* 372. doi:10.1098/rsta.2013.0315

Damjanov, Katarina. (2013). "Lunar cemetery: Global heterotropia and the biopolitics of death." *Leonardo* 46:159–162.

Davies, William (ed.). (2018). *Economic Science Fictions*. Goldsmiths Press.

Davis, Mike. (2000). *Late Victorian Holocausts: El Niño Famines and the Making of the Third World*. Verso.

Day, Dwayne A. (2007). "Take off and nuke the site from orbit. (It's the only way to be sure . . .)." *Space Review*, June 4th.

De Monchaux, Nicholas. (2011). *Spacesuit: Fashioning Apollo*. MIT Press.

Dick, Steven J. (1984). *Plurality of Worlds. The Origins of the Extraterrestrial Life Debate from Democritus to Kant*. Cambridge University Press.

Doctorow, Cory. (2014). "The man who sold the Moon." In Finn, Ed, and Cramer, Kathryn (eds.), *Hieroglyph: Stories and Visions for a Better Future*. William Morrow.

Douglas, Mary. (1966). *Purity and Danger: An Analysis of Concepts of Pollution and Taboo*. Routledge and Keegan Paul.

Elvis, Martin, Milligan, Tony, and Krolikowski, Alanna. (2016). "The Peaks of Eternal Light: A near-term property issue on the Moon." *Space Policy* 38:30–38.

Franklin, Daniel. (2016). "The fab one." *The Economist*, July 21st.

Franklin, H. Bruce. (1980). *Robert A Heinlein: America as Science Fiction*. Oxford University Press.

Galilei, Galileo. (1610). *Sidereus Nuncius*.

Gilbert, Grove Karl. (1898). "The Moon's face: A study of the origin of its features." *Bulletin of the Philosophical Society of Washington*, January.

Godwin, Francis. (1638). *The Man in the Moone or the Discourse of a Voyage thither by Domingo Gonsales*.

Godwin, Robert. (2008). *The Lunar Exploration Scrapbook: A Pictorial History of Lunar Vehicles*. Apogee Books.

Goldblatt, C., Zahnie, K. J., Sleep, Norma H., and Nisbet, E. G. (2010). "The eons of Chaos and Hades." *Solid Earth* 1:1–3.

Grinspoon, David. (2016). *The Earth in Human Hands*. Grand Central Publishing.

Grove, Richard. (1995). *Green Imperialism: Colonial Expansion, Tropical Island Edens and the Origins of Environmentalism, 1600–1860*. Cambridge University Press.

Harland, David M. (2008). *Exploring the Moon: The Apollo Expeditions*. Springer.

Hartmann, William J. (1981). "Discovery of multi-ring basins: Gestalt perception in planetary science." In Schultz, P. H, and Merrill R. B. (eds.), *Multi-Ring Basins: Proceedings of a Luna and Planetary Science Symposium*.

Heinlein, Robert A. (1947). *Rocket Ship Galileo*. G. P. Putnam's Sons.

———. (1950). "The man who sold the Moon." In *The Man Who Sold the Moon*. Shasta Publishers.

———. (1957). "The menace from Earth." *Magazine of Fantasy and Science Fiction*, August.

———. (1966). *The Moon Is a Harsh Mistress*. G. P. Putnam's Sons.

Hubbard, L. Ron, as Northorp, B. A. (1947). "Fortress in the sky." *Air Trails*, May.

Jones, Duncan. (2009). "Moon." Stage 6.

Jones, Eric, and Glover, Ken. (ongoing). *Apollo Lunar Surface Journal*. https://www.hq.nasa.gov/alsj/

Kaku, T., Haruyama, J., Miyake, W., Kumamoto, A., Ishiyama, K., Nishibori, T., Yamamoto, K., Crites, Sarah T., Michikami, T., Yokota, Y.,

Sood, R., Melosh, H. J., Chappaz, L., and Howell, K. C. (2017). "Detection of intact lava tubes at Marius Hills on the Moon by SELENE. (*Kaguya*) lunar radar sounder." *Geophysical Research Letters* 44. doi:10 .1002/ 2017GL074998

Kelly, Thomas J. (2001). *Moon Lander: How We Developed the Apollo Lunar Module.* Smithsonian.

Kepler, Johannes. (1634). *Somnium.*

Kessel, John. (2017). *The Moon and the Other.* Saga Books.

Klerkx, Greg. (2004). *Lost in Space: The Fall of NASA and the Dream of a New Space Age.* Pantheon.

Koeberl, Christian. (2001). "Craters on the Moon from Galileo to Wegener: A short history of the impact hypothesis, and implications for the study of terrestrial impact craters." *Earth Moon and Planets* 85–86:209–224.

Kopal, Zdenek, and Carder, Robert W. (1974). *Mapping of the Moon: Past and Present.* D. Reidel.

Kubrick, Stanley. (1968). "2001: A Space Odyssey." MGM.

Landis, Geoff. (1991). "A walk in the sun." *Isaac Asimov's Science Fiction Magazine,* October.

Launius, Roger D. (1994). *Apollo: A Retrospective Analysis.* NASA.

Laurberg, Marie, Andersen, Anja C., Petersen, Stephen, and Krupp, E. C. (2018). *The Moon—From Inner Worlds to Outer Space*, edited by Lærke Jørgensen. Louisiana Museum of Modern Art.

Lazier, Benjamin. (2011). "Earthrise; or, The Globalization of the World Picture." *American Historical Review,* June, 602–630.

Lewis, H. A. G. (1969). *The Times Atlas of the Moon.* Times Newspaper Publishing.

Lewis, Simon, and Maslin, Mark A. (2018). *The Human Planet: How We Created the Anthropocene.* Pelican Books.

Light, Michael. (1999). *Full Moon.* Jonathan Cape.

Lock, Simon J., and Stewart, Sarah. (2017). "The structure of terrestrial bodies: Impact heating, corotation limits, and synestias." *JGR Planets.* doi:10.1002/2016JE005239

Lock, Simon J., Stewart, Sarah T., Petaev, Michail I., Leinhardt, Zoe M., Mace, Mia T., Jacobsen, Stein B., and Ćuk, Matija. (2018). "The origin of the Moon within a terrestrial synestia." *JGR Planets.* doi:10.1002/2017JE005333

Lockwood, Mike. (2007). "Fly me to the Moon?" *Nature Physics* 3:669–671.

Lofting, Hugh. (1928). *Doctor Dolittle in the Moon.* Frederick A. Stokes.

Logsdon, John M. (2013). *John F. Kennedy and the Race to the Moon.* Palgrave Macmillan.

———. (2014). "John F. Kennedy and the 'Right Stuff.'" *Quest* 20:4–15.

Lovelock, James. (1979). *Gaia: A New Look at Life on Earth.* Oxford University Press.

Lubin, Philip. (2016). "A roadmap to interstellar flight." *Journal of the British Interplanetary Society* 69:40–72.

Lunar Exploration Analysis Group. (2016). *Exploring the Moon in the 21st Century: Themes, Goals, Objectives, Investigations, and Priorities.* https://www.lpi.usra.edu/leag/

MacDonald, Alexander. (2017). *The Long Space Age: The Economic Origins of Space Exploration from Colonial America to the Cold War.* Yale University Press.

MacKay, Angus. (1971). *Super Nova and the Frozen Man.* Knight Books.

Maher, Neil M. (2015). *Apollo in the Age of Aquarius.* Harvard University Press.

Mailer, Norman. (1971). *Of a Fire on the Moon.* Pan Books.

Marvin, Ursula B. (1986). "Meteorites, the Moon and the history of geology." *Journal of Geological Education* 34:140–165.

McCray, W. Patrick. (2012). *The Visioneers: How a Group of Elite Scientists Pursued Space Colonies, Nanotechnologies, and a Limitless Future.* Princeton University Press.

McDonald, Ian. (2015). *Luna: New Moon.* Tor Books.

———. (2017). *Luna: Wolf Moon.* Tor Books.

McDougall, Walter A. (1985). *The Heavens and the Earth: A Political History of the Space Age.* Basic Books.

Mendlesohn, Farah. (2019). *The Pleasant Profession of Robert A. Heinlein.* Unbound.

Metzger, Philip T., Muscatello, A., Mueller, R. P. and Mantovani, J. (2013). "Affordable, rapid bootstrapping of space industry and solar system civilization." *Journal of Aerospace Engineering* 26:18–29.

Miller, Charles, Wilhite, Alan, Cheuvront, Dave, Kelso, Rob, McCurdy, Howard, and Zapata, Edgar. (2015). "Economic assessment and systems analysis of an evolvable lunar architecture that leverages commercial space capabilities and public-private-partnerships." NexGen Space LLC.

Miller, Walter M., Jr. (1957). "The lineman." *Magazine of Fantasy and Science Fiction*, August.

Mindell, David A. (2008). *Digital Apollo: Human and Machine in Spaceflight.* MIT Press.

Montgomery, Scott L. (1999). *The Moon and the Western Imagination.* University of the Arizona Press.

Moore, C. L. (1936). "Lost paradise." *Weird Tales,* July.

Moore, Jason. (2015). *Capitalism in the Web of Life: Ecology and the Accumulation of Capital.* Verso.

Morton, Oliver. (1999). "Looking for life." *Newsweek International.*

———. (2016). "A sudden light." *The Economist,* September 1st.

Musk, Elon. (2018). "Making life multi-planetary." *New Space* 6:2–11.

NASA. (1969). *Apollo 8 Onboard Voice Transcription.* NASA.

Nasmyth, James. (1882). *James Nasmyth: Engineer; an Autobiography,* edited by Samuel Smiles. John Murray.

Nasmyth, James, and Carpenter, James. (1871). *The Moon: Considered as a Planet, a World, and a Satellite.* James Murray.

National Academy of Sciences. (2007). *The Scientific Context for Exploration of the Moon.* National Academies Press.

Nevala-Lee, Alec. (2018). *Astounding: John W. Campbell, Isaac Asimov, Robert A. Heinlein, L. Ron Hubbard, and the Golden Age of Science Fiction.* Dey Street Books.

Niven, Larry. (1976). *A World Out of Time.* Holt, Rinehart and Winston.

———. (1980). *The Patchwork Girl.* Ace Books.

Niven, Larry, and Pournelle, Jerry. (1977). *Lucifer's Hammer.* Playboy Press.

Nye, David E. (2003). *America as Second Creation: Technology and Narratives of New Beginnings.* MIT Press.

Oberth, Herman. (1923). *Die Rakete zu den Planetenräumen.* R. Oldenbourg.

Olson, Valerie, and Messeri, Lisa. (2015). "Beyond the Anthropocene: Un-Earthing and Epoch." *Environment and Society: Advances in Research* 6:28–47.

O'Neill, Gerard K. (1976). *The High Frontier: Human Colonies in Space.* William Morrow.

Pal, George. (1950). "Destination Moon." George Pal Productions.

Parker, Geoffrey. (2014). *Global Crisis: War, Climate Change and Catastrophe in the Seventeenth Century.* Yale University Press.

Patterson, William H., Jr. (2011). *Robert A. Heinlein: In Dialogue with His Century: Volume 1: 1907–1948, Learning Curve.* Tor Books.

———. (2016). *Robert A. Heinlein: In Dialogue with His Century: Volume 2: 1948–1988, The Man Who Learned Better.* Tor Books.

Perry, William. (2015). *My Journey at the Nuclear Brink*. Stanford Security Studies.

Poole, Robert. (2008). *Earthrise: How Man First Saw the Earth*. Yale University Press.

Potter, Michael. (2008). "Orphans of Apollo." Free Radical Productions.

Pournelle, Jerry. (1981). *A Step Farther Out*. Baen Books.

Pumfrey, Stephen. (2011). "The Selenographia of William Gilbert: His pre-telescopic map of the Moon and his discovery of lunar libration." *Journal for the History of Astronomy* xlii:1–11.

Reeves, Eileen. (1997). *Painting the Heavens: Art and Science in the Age of Galileo*. Princeton University Press.

Riccioli, Giovanni Battista. (1651). *Almagestum Novum*.

Richardson, Robert S. (ed.). (1961). *Man and the Moon*. World Publishing.

Riley, Christopher. (2009). *Apollo 11 Manual*. Haynes.

Roberts, Adam. (2016). *The History of Science Fiction*. 2nd ed. Palgrave Macmillan.

Robertson, Frances. (2006). "James Nasmyth's photographic images of the Moon." *Victorian Studies* 48:595–693.

Robinson, Kim Stanley. (2018). *Red Moon*. Orbit.

Roy, Arpita, Wright, Jason T., and Sigurdsson, Stein. (2014). "Earthshine on a young Moon: Explaining the lunar farside highlands." *Astrophysical Journal Letters* 788:L42.

Scott, David Meerman, and Jurek, Richard. (2014). *Marketing the Moon: The Selling of the Apollo Lunar Program*. MIT Press.

Scott, Zack. (2017). *Apollo: The Extraordinary Visual History of the Iconic Space Programme*. Wildfire.

Serviss, Garrett P. (1898). "Edison's conquest of Mars." *New York Evening Journal*, January and February.

Silk, Joseph. (2018). "Put telescopes on the far side of the Moon." *Nature* 553:6.

Simak, Clifford. (1960). "The trouble with Tycho." *Amazing Stories*, October.

Sleep, Norman H., and Zahnle, Kevin. (1998). "Refugia from asteroid impacts on early Mars and the early Earth." *Journal of Geophysical Research* 103:28,528–28,529, 28,544.

Sobel, Dava. (2005). *The Planets*. Fourth Estate.

Spudis, Paul D. (1996). *The Once and Future Moon*. Smithsonian Institution Press.

———. (2016). *The Value of the Moon: How to Explore, Live, and Prosper in Space Using the Moon's Resources.* Smithsonian Institution Press.

Spudis, Paul D., and Richards, Robert. (2018). "Mission to the Rima Bode Regional Pyroclastic Deposit." Presentation at the Lunar Science for Landed Missions workshop, NASA Ames Research Center, January.

St Clair, Kassia. (2018). *The Golden Thread: How Fabric Changed History.* John Murray.

Stephenson, Neal. (2015). *Seveneves.* HarperCollins.

Sterzick, Michael F., Bagnulo, Stefano, and Palle, Enric. (2011). "Biosignatures as revealed by spectropolarimetry of Earthshine." *Nature* 483:64–66.

Stubbs, Timothy, Collier, Michael, Farrell, Bill, Keller, John, Espley, Jared, Mesarch, Michael, Chai, Dean, Choi, Michael, Vondrak, Richard, Purucker, Michael, Malphrus, Ben, Zucherman, Aaron, Hoyt, Robert, Tsay, Michael, Halekas, Jasper, Johnson, Tom, Clark, Pam, Kramer, Georgiana, Glenar, Dave, and Gruesbeck, Jacob. (2018). "Bi-Sat Observations of the Lunar Environment Above Swirls (BOLAS): Tethered microsat investigation of space weathering and the water cycle at the Moon." Paper presented at the *49th Lunar and Planetary Science Conference*, abstract 2394.

Swanwick, Michael. (1992). *Griffin's Egg.* Legend.

Tennyson, Alfred Lord. (1842). "Locksley Hall." In *Poems.* Moxon.

Tumlinson, Rick N., with Medlicott, Erin R. (eds.). (2005). *Return to the Moon.* Apogee Books.

Vance, Ashlee. (2015). *Elon Musk: How the Billionaire CEO of SpaceX and Tesla Is Shaping Our Future.* Virgin Books.

Varley, John. (1992). *Steel Beach.* Ace Books.

Verne, Jules. (1865). *De la Terre a la Lune.*

———. (1870). *Autour de la Lune.*

Von Trier, Lars. (2011). "Melancholia." Zentropa.

Waltham, David. (2016). *Lucky Planet: Why Earth Is Exceptional—and What That Means for Life in the Universe.* Icon Books.

Weir, Andy. (2017). *Artemis.* Del Rey.

Wells, H. G. (1898). *The War of the Worlds.* William Heinemann.

———. (1901). *The First Men in the Moon.* Bowen-Merrill.

———. (1902). *The Discovery of the Future.* Fisher Unwin.

Westwick, Peter J. (2018). "From the Club of Rome to Star Wars: The era of limits, space colonization and the origins of SDI." In Geppert, Alexander (ed.), *Limiting Outer Space: Astroculture After Apollo.* Springer.

Whitaker, Ewen A. (2008). *Mapping and Naming the Moon: A History of Lunar Cartography and Nomenclature.* Cambridge University Press.

Wilhelms, Don A. (1993). *To a Rocky Moon: A Geologist's History of Lunar Exploration.* University of Arizona Press.

Wilkins, John. (1638). *The Discovery of a World in the Moone.*

Williamson, Jack, and Breuer, Michael. (1931). "The Birth of a New Republic." *Amazing Stories Quarterly.*

Wingo, Dennis. (2004). *Moonrush: Improving Life on Earth with the Moon's Resources.* Apogee Books.

———. (2016). "Site selection for lunar industrialization, economic development, and settlement." *New Space* 4:19–39.

Wood, Gillen D'Arcy. (2014). *Tambora: The Eruption That Changed the World.* Princeton University Press.

Woods, David. (2016). *NASA Saturn V Manual.* Haynes.

Woods, David, with others. (ongoing). *Apollo Flight Journal.* https://history.nasa.gov/afj/

Woolf, N. J., Smith, P. S., Traub, W. A., and Jucks, K. W. (2002). "The spectrum of earthshine: A pale blue dot observed from the ground." *Astrophysical Journal* 574:430–433.

Zubrin, Robert. (2018). "Moon direct." *New Atlantis,* October 31st.

INDEX

OLIVER MORTON has written about space for decades in publications ranging from *The Economist* to *Nature* to *Prospect* to the *New Yorker*. Asteroid 10716 olivermorton is named after him.

PublicAffairs is a publishing house founded in 1997. It is a tribute to the standards, values, and flair of three persons who have served as mentors to countless reporters, writers, editors, and book people of all kinds, including me.

I. F. STONE, proprietor of *I. F. Stone's Weekly*, combined a commitment to the First Amendment with entrepreneurial zeal and reporting skill and became one of the great independent journalists in American history. At the age of eighty, Izzy published *The Trial of Socrates*, which was a national bestseller. He wrote the book after he taught himself ancient Greek.

BENJAMIN C. BRADLEE was for nearly thirty years the charismatic editorial leader of *The Washington Post*. It was Ben who gave the *Post* the range and courage to pursue such historic issues as Watergate. He supported his reporters with a tenacity that made them fearless and it is no accident that so many became authors of influential, best-selling books.

ROBERT L. BERNSTEIN, the chief executive of Random House for more than a quarter century, guided one of the nation's premier publishing houses. Bob was personally responsible for many books of political dissent and argument that challenged tyranny around the globe. He is also the founder and longtime chair of Human Rights Watch, one of the most respected human rights organizations in the world.

•　　　•　　　•

For fifty years, the banner of Public Affairs Press was carried by its owner Morris B. Schnapper, who published Gandhi, Nasser, Toynbee, Truman, and about 1,500 other authors. In 1983, Schnapper was described by *The Washington Post* as "a redoubtable gadfly." His legacy will endure in the books to come.

Peter Osnos, *Founder*